BIRDS
PHILIPPINES

POCKET PHOTO GUIDES

Tim Fisher
Photographs by Nigel Hicks

H E L M
LONDON · OXFORD · NEW YORK · NEW DELHI · SYDNEY

HELM
Bloomsbury Publishing Plc
50 Bedford Square, London, WC1B 3DP, UK

BLOOMSBURY, HELM and the HELM logo are trademarks of
Bloomsbury Publishing Plc

First published by New Holland in 2001
This edition first published by Bloomsbury, 2020

Copyright © Tom Fisher, 2020
Photographs © Nigel Hicks, 2020 (except where otherwise listed
on page 144)

Tom Fisher has asserted his right under the Copyright, Designs and Patents
Act 1988 to be identified as the author of this work

All rights reserved. No part of this publication may be reproduced or
transmitted in any form or by any means, electronic or mechanical,
including photocopying, recording, or any information storage or retrieval
system, without prior permission in writing from the publishers
Bloomsbury Publishing Plc does not have any control over, or
responsibility for, any third-party websites referred to or in this book. All
Internet addresses given in this book were correct at the time of going
to press. The author and publisher regret any inconvenience caused if
addresses have changed or sites have ceased to exist, but can accept no
responsibility for any such changes

A catalogue record for this book is available from the British Library

Library of Congress Cataloguing-in-Publication data has been applied for

ISBN: 978-1-4729-8254-4

2 4 6 8 10 9 7 5 3 1

Designed by Susan McIntyre

Printed and bound in India by Replika Press Pvt. Ltd.

To find out more about our authors and books, visit www.bloomsbury.com
and sign up for our newsletters

CONTENTS

The Philippine Islands	4
How to use this book	5
Glossary	7
Birds and habitats	8
When to visit	10
Field techniques	11
Where to go birdwatching	13
Map of the Philippine Islands	13
Species descriptions	20
Index	141
Photo credits	144

THE PHILIPPINE ISLANDS

The Philippine Islands are a cluster of over 7,000 islands, two-thirds of which are little more than rocks and reefs and only 470 more than a mile square. They extend 1,800km from just south of Taiwan and lie north of Sulawesi and east of Borneo, forming the eastern perimeter of South-East Asia. The total area of the country is 300,000 square kilometres, slightly larger than the United Kingdom, of which two-thirds is made up by the two largest islands, Mindanao and Luzon. The next largest islands are Samar, Negros, Palawan, Panay and Mindoro, ranging from 10,000 to 13,200 square kilometres.

Some of the islands, particularly Luzon and Mindanao, have major mountain ranges, with peaks reaching nearly 3,000m; these areas have a different climate and vegetation to the lowlands. Overall, the Philippines has a varied rugged terrain with numerous bays and gulfs that reflect the complex origins of the archipelago. The large number of islands, with their different origins and altitudinal ranges, plus the warm humid climate, plentiful rain and rich soil, has resulted in a considerable range in biodiversity. There are 12,000 species of plants, of which 8,500 are flowering types, with 3,800 species of native tree. There are more than 165 species of mammals, 270 species of reptiles and amphibians and over 560 species of birds, of which more than 170 are endemic to the Philippines.

No doubt practically the whole archipelago was covered in virgin forest at one point in time. Much was cleared over the centuries for farming and fuel, particularly in well-populated areas such as central Luzon, Cebu and Bohol. Early last century there was large-scale clearing for sugar plantations, notably on Negros. From the 1960s to the 1980s, under the Marcos regime, much of the remaining forest was divided up into timber concessions whose access roads allowed the kaingineros (slash-and-burn farmers) to move in and clear the land for subsistence farming. These practices, along with the increase in population from 10 million in 1920, 52 million in 1982 and 89 million in 2006, and the consequent demand for land, have led to the clearance of nearly all lowland forest throughout the country, just leaving patches on the mountain tops. Even the national parks, established to conserve the natural forest and its wildlife, have suffered from exploitation of the timber and encroaching farmers. A large number of Philippine birds are now severely threatened due to the loss of their habitat and are in danger of becoming extinct. The conservation of remaining patches of forest on all islands should be of the highest priority so as to save the islands' marvellous and varied biodiversity.

This book is an introduction to the birds of the Philippines. Out of a total of over 600 species recorded in the country the book has illustrated, with photographs, 215 of them. The birds of the Philippines are not well known and for many species this is probably the first time a photograph has been published. As far as possible a cross section of the families has been shown so that the beginner can appreciate the variety of birds found here. Overall, the book has concentrated on those commoner species that are more likely to be seen by the average birdwatcher, particularly those birds found on the outskirts of Metro

Manila and in the open country and coastal areas. There are, however, many species that are restricted to the undergrowth of the forest. Owing to the extreme difficulty of photographing these birds in the dark understorey, many have been photographed in the hand so that the reader can see their diagnostic features more readily. This book is a useful introduction to the birds of the Philippines and it will hopefully stimulate interest in the subject.

HOW TO USE THIS BOOK

This book has been designed with clarity and ease of use in mind. The photographs show the commonly seen plumage. Sometimes there are two photographs which usually show two different forms or subspecies of the same species. The English names of the species are the generally accepted ones and are in accordance with *Birdwatching in the Philippines* (Volume 1).

The species descriptions begin with the English name followed by the scientific name and the overall size of each bird. The rest of the text indicates both its distribution worldwide and within the Philippines along with its scarcity, whether it is a resident or a migrant, and its preferred habitat. Second, the diagnostic features of its plumage, habits and calls are given to enable the observer to identify the bird.

The 600 or so species of Philippine birds are divided into many different subspecies between the islands, the total reaching nearly one thousand. The great majority are forest birds and include such families as hornbills, parrots, pigeons, cuckoos, pittas, woodpeckers, babblers, bulbuls, flycatchers, sunbirds and flowerpeckers. One family is endemic to the Philippines, the Rhabdornithidae Philippine creepers, which comprises three species. A number of species are restricted to one island only, and on these islands are restricted to certain mountain tops. In fact, owing to the remoteness of some areas, the Philippines is one of the most likely places on earth for there still to be forms as yet unknown to science. In the past 25 years, there have been at least three new species discovered and many new subspecies recognized, whilst further fieldwork will undoubtedly show that different forms of one species are found to be separate species in their own right.

Many species are threatened with extinction owing to loss and destruction of habitat. One of the most notable birds under threat is the Philippine Eagle *Pithecophaga jefferyi*, which may be down to a population of between 200 and 300 birds. A pair will hold territory in a large tract of forest, but loss of extensive forest cover has severely threatened the bird. The fact that it only breeds every other year, rearing only a single chick, only makes its plight all the more critical. Some islands are practically devoid of forest cover altogether, Cebu being the prime example. It has two of its own endemics, the Cebu Black Shama *Copsychus cebuensis* and the Cebu Flowerpecker *Dicaeum quadricolor*, along with several endemic subspecies. The flowerpecker was thought to be extinct since 1906, but the discovery of a tiny remnant patch of forest in the central highlands led to its rediscovery in 1992. The pressure on the birds from loss of habitat continues practically unabated today mainly as a result of the continuous demand for land for agriculture.

Birds in the Philippines are also subject to intensive hunting pressure, particularly those good to eat such as large pigeons and hornbills; in fact, large populations of pigeons and hornbills are a clear sign that the forest is in a healthy state. Many types of birds are trapped for the pet trade, parrots in particular, the most notorious case being the Philippine Cockatoo *Cacatua haematuropygia*, which was common over most of the Philippines but now only reliably to be found on Palawan and the Sulu Islands, where it is rare and still under threat. In the mountain regions of northern Luzon, birds have traditionally been trapped in large numbers at night by the use of bright lights. Dalton Pass in Nueva Vizcaya was a well-known location for this, where not only migrants from overseas were caught, but also many resident species once thought to be sedentary but now known to undertake local migrations. At least two species, Worcester's Buttonquail *Turnix worcesteri* and Brown-banded or Luzon Rail *Lewinia mirifica*, were trapped regularly there even though no ornithologist has seen them alive in the field nor knows where they breed. The migratory duck which visit the Philippines every winter between October and April and which can be seen in such numbers at Candaba Marsh are subject to heavy shooting from 'well-to-do' sportsmen.

In recent years, there has been some action to stem the losses. Commercial logging has practically ceased and a total log ban is being considered, but government and non-government organizations will have to work at the grassroots level with local communities to prevent any further destruction of the forest cover. New reserves are being gazetted and others being considered, whilst tree plantations for timber and fuel are being established in order to save the remaining tracts. There are several organizations dedicated to conservation. Probably the most important is the Haribon Foundation, which also runs the local ornithological society.

GLOSSARY

Axillaries The feathers in the wing-lining.

Carpal joint The main bend of the wing.

Casque A decorative enlargement of the upper mandible, typically in hornbills.

Cere Bare skin around the base of the bill.

Crepuscular Active in twilight and just before dawn.

Crest A tuft of elongated feathers on the crown that can often be raised and lowered.

Crown The top of the head, often distinctly coloured; not so extensive as the cap.

Diagnostic A character of sufficient distinctiveness to allow identification to be made.

Endemic Indigenous; restricted to a particular area.

Flight feathers The primaries and secondaries.

Frontal shield Bare horny or fleshy skin on the forehead.

Gular pouch Expandable piece of skin on throat of frigatebirds.

Lores Area between the base of the bill and the eye.

Malar area The area bounded by base of bill, side of throat and below eye.

Mantle The back, upper wing-coverts and scapulars combined.

Migrant A bird travelling between its breeding area and wintering area.

Nape The back of head and back of neck.

Ocelli Eye-like spots on the plumage.

Primaries The flight feathers on the outer joint of the wing.

Primary forest Original or virgin forest.

Resident A bird that stays all year round in the country and breeds.

Scapulars Feathers over the shoulder blades.

Secondary forest New forest growing where primary forest had been.

Shaft-streak A streak made when the shaft is a different colour to the feather.

Subterminal Usually refers to band near end of tail or on wings.

Tail-coverts Small feathers above and below the tail covering the base.

Wing-coverts Small feathers covering the base of the primaries and secondaries.

BIRDS AND HABITATS

The tropical island habitats with forested lowlands and mountain ranges, along with areas of swamp and marshland and long coastline, are very favourable to a rich and diverse birdlife. At one time, virtually the whole country was covered in forest from the coast to the mountain tops. Even the flooded plains were forested. Today man has changed all that. Little remains of the original forest and in its place are cultivations, grasslands, towns, open swamps and plantations amongst other habitats. The range of habitats can be divided as follows.

Open sea and coastline
Tidal mudflats and reefs, with their abundant food supplies, are the feeding grounds for a variety of shorebirds, egrets and herons, whilst terns and boobies are found in the open sea and breeding on remote rocky islets.

Coastal vegetation, mangroves and estuaries
Mangrove forest occupies intertidal areas and, although few use it as their principal habitat, it provides food and sanctuary for many birds, such as Golden-bellied Gerygone *Gerygone sulphurea*, Pied Fantail *Rhipidura javanica*, Olive-backed Sunbird *Cinnyris jugularis* and Yellow-vented Bulbul *Pycnonotus goiaver*. Cleared mangroves with resultant fishponds are attractive to herons and waders. Sandy shores are the principal habitat of the Malaysian Plover *Charadrius peronii*.

Rivers, lakes and marshlands
Large rivers with vegetated banks offer similar habitat to wetlands whilst small rivers, particularly those in forests with small fish and frogs, are the haunt of a variety of birds such as kingfishers, small herons, rails and birds of prey such as the Chinese Goshawk *Accipiter soloensis*. Small marshes and ponds are home to White-breasted Waterhen *Amaurornis phoenicurus*, grassbirds and munias. There are few remaining large tracts of marshland, but those that do exist, such as at Candaba and Agusan, are excellent for a large variety of wetland birds, including herons, egrets, harriers, jacanas, snipe, rails and warblers, plus a host of winter visitors such as several species of duck and waders. The characteristic vegetation includes low-growing marsh plants, reedbeds, water hyacinth and a variety of palms. Even overgrown disused fishponds and wet paddies can hold a number of water birds.

Forests
Most of the Philippines was once covered in forest. Lowland dipterocarp forest lay almost immediately inland from the coastal zone and was or still is followed by mid-montane forests and montane forests. They are the richest habitat in terms of the number of species and individuals of birds. Most of the Philippines' endemic birds are forest-dwellers. The lowland forests are noted for their tall canopy and three tiers of trees, tall shrubs and undergrowth. Birds inhabit all sectors, with pigeons and hornbills in the tall canopy, flycatchers and babblers and members of mixed feeding flocks in the mid-storey, and ground-feeding birds such as

babblers and pittas. Mid-montane forests are found on slopes at 900–1,500m. They are dense and medium-height, and possess a wealth of undergrowth consisting of ferns, lianas and epiphytes. This forest type is rich in birds that are often different from those found in the lowlands. In the high mountains dense mossy oak forests occur, along with a host of specialized montane birds. On Luzon, on some slopes, there are pine forests, the habitat of the Red Crossbill *Loxia curvirostra*.

The forest edge and secondary forests, if not disturbed, can be good for birds, particularly those hard-to-see canopy birds, whilst logged-over areas with scattered trees are favoured by such birds as Dollarbird *Eurystomus orientalis*, Coleto *Sarcops calvus* and Philippine Falconet *Microhierax erythrogenys*.

Grassland and scrubland

Owing to man's destruction of the forest, there are extensive areas of cogonal grass in the foothills which get regularly burnt. If areas are left alone and conditions become favourable, other grasses and shrubs grow, creating an open scrubland habitat. There are a number of birds that favour this habitat and which have become commoner as a result, such as Tawny Grassbird *Megalurus timoriensis*, Grass Owl *Tyto capensis*, Long-tailed Shrike *Lanius schach* and Pied Bushchat *Saxicola caprata*. When adjacent to original forests, the habitat can be even richer in birds.

Cultivated areas, villages and towns

With the advent of man and his agricultural practices, new habitats have grown up. Rice fields are good for waders, munias and doves when unplanted. Coconut groves, orchards, gardens and bamboo thickets host a number of species such as Oriental Magpie-robin *Copsychus saularis*, Pied Fantail *Rhipidura javanica*, Lowland White-eye *Zosterops meyeni* and Golden-bellied Gerygone amongst many others. Even the rare and endangered Cebu Black Shama has survived in the small bamboo thickets just outside the island's capital city. In the gardens in towns can be found Brown Shrikes *Lanius cristatus*, Olive-backed Sunbird and Yellow-vented Bulbul, whilst the buildings offer nesting sites for Tree Sparrows *Passer montanus* and Striated Swallows *Cecropis striolata*.

WHEN TO VISIT

In planning a birdwatching visit to the Philippines or a day out to watch birds if you are resident, many things should be taken into account. These would include what is the best time of year climate-wise, which islands to visit, where to go, what field equipment and clothing to take, how to behave and how to go about looking for and identifying birds in the field.

The climate not only changes at different times of the year, it also varies between islands and altitudes. Although it may be hot and dry in Luzon, Mindanao may well be wet, whilst a visit to the central Cordilleras can be quite cold, particularly at night and in the early morning. Rainfall is the most important climatic element, as the Philippines lies in the intertropical convergence zone. Typhoons (cyclones) initiate off the Marianas and regularly hit the eastern coast of Luzon and Samar between the months of July and November. The seasonality of the rainfall can be generally divided as follows:

Type 1: November to April dry, rest of the year wet (this prevails in the western parts of Luzon, Mindoro, Panay, Negros and Palawan);

Type 2: no true dry season but highest rainfall from November to January (southern Luzon, Samar, Leyte and eastern Mindanao).

Elsewhere it is not so pronounced but generally drier from November to April.

FIELD TECHNIQUES

Birdwatching is not only an absorbing hobby but, when the observer makes detailed notes of their observations, their results are important to scientists in monitoring changes in the environment. The following are a few pointers to assist birdwatchers in making their observations accurate and useful.

Field equipment Probably the most important piece of equipment is a pair of binoculars. The ideal size is probably 8 × 40 which has a relatively wide aperture, allowing in a lot of light which is particularly important when observing birds in the dark understorey of the forest. They should also be fairly light in weight. Telescopes, more pertinent for experienced birders, are useful for identifying birds at a distance, such as on estuaries, and for seeing them up close, but are difficult to use in the forest. It is always a good idea to carry a field guide in case of new or little-known birds. A quick reference will indicate an important diagnostic feature one might otherwise have missed. Also, a notebook and pencil or ballpen are essential to take down details of your observations. All the above, plus your camera, should be kept in waterproof bags, when not in use, to protect them from the ever-present rain and humidity. It is a good idea to carry a clean handkerchief or cloth to dry the lenses on the binoculars. Clothing should be lightweight and drab, such as brown or green, not red or white. Protection from the rain is important; a small folding umbrella is very useful. A plastic sheet to sit on when on wet and dirty ground is a good idea.

Behaviour in the field When birdwatching in the forest, or indeed anywhere, it is important to keep conversation to a minimum and to avoid sudden movements. Do not suddenly point at or rush towards a newly observed bird; they are very sensitive to sound and movement and will fly away immediately. It is best to keep to the trails to avoid snapping twigs and branches; it is also a good idea to use a tree as camouflage when standing still. When remaining motionless, sit down as comfortably as possible, to avoid later movement, and make yourself as obscure as possible. Generally walking slowly works best, although a fast silent walk along a clean trail can be an excellent way to come across skulkers such as the Palawan Peacock-pheasant *Polyplectron napoleonis* or pittas. Pishing or squeaking can be a useful way to attract birds like flycatchers and babblers, whilst tape playback of calls or songs can be an excellent way of observing skulking birds that would be difficult to see otherwise.

Birdwatching usually takes the observer to places away from popular tourist destinations. When visiting national parks and reserves, it is important to ascertain if a permit is required, and obtain it beforehand if it is. Many places will be privately owned so it is a good idea to obtain permission from the owner if possible and also, especially in more remote areas and ancestral domains, to meet with the local officials such as the mayor, barangay chief or village headman. Always leave an area as you found it.

Dawn is the best time to watch birds in the field as then they are noisy and active. In fact it is a good idea to be there before dawn as this is the time, often the only time, that the large forest kingfishers and owls such as the Philippine Scops-owl *Otus megalotis* can be seen. Bird activity dies off by midday and picks up again in the late afternoon but never to the same degree as at dawn. However, if the day is cloudy and overcast, birds often stay fairly active throughout. A heavy shower followed by sunshine often stimulates a burst of activity.

Keeping field notes Birdlife in the Philippines is very little understood. For example, the breeding habits of around fifty percent of the resident species are not known, and the situation is similar for the songs and calls. Our knowledge of the distribution is incomplete and species are regularly being found in new localities and on new islands. It is therefore essential, if your observations are to be of use to science, that you note down details of your observations whilst in the field. A prepared checklist with columns for numbers seen is useful and time-saving, besides making one aware of birds that have still not been seen during the day. In making notes, the following pointers should be useful, all of which should be done in the field and not later at home.

(1) Put down the name of the locality and the type of habitat in the area, altitude, weather conditions and time and date of visit.
(2) Record all birds seen at each locality.
(3) Make detailed notes of rare or little-known birds which can be later confirmed by referring to the literature. It is important to send notes of rare birds to the local associations, such as the Haribon Foundation, with as full details as possible. Besides those points in (1), also list the names of all observers, the distance and observation conditions, including optics used, a full description of the bird along with notes on its behaviour and any other pertinent facts such as prior experience of the species or unusual weather conditions.
(4) Keep notes of all breeding birds whether feeding young or sitting on the nest along with the type of nest used and its location.
(5) Record all unusual behaviour.

Identification hints Correct recognition of a type of bird is a combination of several factors, including its appearance, voice and behaviour. Note down as much of the bird's plumage colour and pattern as possible and not just the most obvious features. There are a number of species that are very similar to each other, sharing some obvious characteristics but whose diagnostic features are more obscure. Make a simple sketch that details the size, shape, length of bill and tail, the presence of a crest or other features, colour of plumage, eyes, bill and feet. Also note down calls, habitat and behaviour, e.g. standing erect on the top of a bush, climbing up the side of a tree trunk or walking on the ground. If a bird closely resembles a common bird, note this along with the variants.

WHERE TO GO BIRDWATCHING

Main Islands
1. Luzon
2. Mindoro
3. Marinduque
4. Burias
5. Tablas
6. Sibuyan
7. Masbate
8. Samar
9. Leyte
10. Panay
11. Palawan
12. Negros
13. Cebu
14. Bohol
15. Mindanao

Birdwatching in the Philippines is difficult and complicated due to the remoteness of many of the sites, the majority of which have no nearby accommodation. This means either travelling perhaps an hour or so to the site or camping nearby. National parks are no exception, with little hands-on protection except at a small minority such as St Paul's Underground River National Park in Palawan. It is well worth the effort, however, because of the variety of beautiful endemic birds that can be found. In order to see a large variety of birds, particularly the endemic species, it is necessary to visit anywhere from ten to fifteen sites in a three- to four-week visit. This would entail visiting the lowland and montane forests and marshlands of Luzon and Mindanao plus several other islands including Palawan, Cebu, Negros and Bohol if one has time. This is due to each island's great bird diversity, many species being restricted to one island only.

LUZON

The fertile central plain, once an extensive area of marshes and floodplains, is now the ricebowl of the country, with only small patches of wetland such as Candaba Marsh. In the north are the high Cordillera range in the centre and the Sierra Madre mountains along the eastern seaboard, reaching as far south as the Angat Watershed. The southern half of Luzon holds fertile volcanic plains with a number of forested mountains. The island has many endemics, some of which are restricted to just a portion of the island. There are many good birdwatching sites, of which only the better known and more accessible are mentioned here.

Mount Makiling

Mount Makiling is situated near the spa town of Los Banos, 50km south of Manila. It is very suitable for a day out birding for Manila residents. The forest is a nature reserve managed by the University of the Philippines. There are over 1,000 hectares of tall dipterocarp forest coming down to an altitude of around 100m, containing a large variety of lowland forest specialities.

Quezon National Park

This park lies 150km south of Manila on the main highway to the south between Lucena and Atimonan in Quezon province. It comprises around 1,000 hectares of original dipterocarp forest growing on limestone rock. Although badly logged in recent years, it still holds a great variety of Luzon's birds including hornbills, pigeons, malkohas and woodpeckers. The nearby fishponds and mangroves just south of Pagbilao are excellent for waterbirds.

Hamut Camp

Good forest is becoming harder and harder to find on Luzon. This area in the northern Sierra Madre is one of the richest remaining areas, but is difficult to reach (arrangements are best made via the author). Access is from Tuguigarao and it is necessary to camp. Almost all the Luzon forest birds can be found here, including a good variety of pigeons, babblers and flycatchers, but the star attraction is the shy but spectacular Whiskered Pitta.

Subic Freeport Zone

Subic used to be an American naval base, and is situated in Bataan province north-west of Manila. The area is vast and contains some of the last pieces of lowland forest remaining in Luzon. Owing to restrictions on public access and its general protection, many birds rare elsewhere are relatively common here, such as Green Imperial-pigeons *Ducula aenea* and Blue-naped Parrots *Tanygnathus lucionensis*. It is about a four-hour drive from Manila and permission is required from the Ecology Centre to go into the old Naval Magazine area where some of the best birding is to be had.

Mount Polis, Mountain Province

Mount Polis is situated north-west and an hour's drive from Banaue, a popular tourist destination and well known for its rice terraces. It is one of the most accessible sites for Luzon mountain endemics, as one can drive up to an altitude of around 2,000m on the Banaue to Bontoc road and watch birds along the roadside. The vegetation is primarily oak forest but some of the slopes are covered in pine forest. The mountain streams are excellent for the Luzon Water-redstart *Rhyacornis bicolor*.

Candaba Marshes, Pampanga

This area is the last remaining piece of marsh in what was once a vast floodplain and swamp lying just north of Manila. Access is by the Baliuag–Candaba and the Candaba–San Miguel roads. It is best between October and December when the area is still very wet and is excellent for wetland birds including a large variety of migratory ducks.

PALAWAN

Palawan is a long narrow island with a steeply sloping central mountain range of moderate height though reaching 2,000m in places. It forms a link between the Philippines and Borneo. There is a distinct Malaysian faunal element in the birds, and many species that occur in Borneo also occur in Palawan. The island holds about 20 Philippine endemic species, of which 14 are entirely restricted to it. Historically, the human population has been low and the resulting pressure on the environment was correspondingly small. Even now the island is still relatively well forested. Ornithologically, the area is not well studied, particularly the islands in the south such as Balabac and in the north such as Busuanga. To see a good proportion of Palawan's birds a visit to the following is recommended.

St Paul's Underground River National Park

The park lies about 100km and a two- to three-hour drive north of Puerto Princesa on the west coast. It comprises about 5,000 hectares and contains towering limestone cliffs and white sandy beaches and is famous for its 8km underground river. The magnificent dipterocarp forest reaches the shore and contains a large variety of Palawan birds, including Palawan Peacock-pheasant, and is probably the best place in the world to see the Philippine Cockatoo.

Iwahig Penal Colony

Situated about half an hour's drive just south of Puerto Princesa, this is an excellent place for lowland birds and in particular the Palawan Flycatcher *Ficedula platenae* and Melodious Babbler *Malacopteron palawanense*, which are difficult to find at St Paul's. The area also contains fishponds and paddies which are good for waders, quail and rails.

Rasa Island

Lying just 1–2km off the town of Narra on the southern coast of Palawan, this is one of the last refuges of Philippine Cockatoo, with over 100 birds sometimes roosting on the island. It also holds Great-billed Heron and

Mantanani Scops Owl *Otus mantananensis*, the latter a small-island specialist confined to the Philippines and a few islands off Sabah. Recently declared a Wildlife Sanctuary.

Ursula Island
Ursula is a 17-hectare forested island lying 18km off the south end of Palawan in the Sulu Sea. It is an excellent site for Nicobar Pigeon *Caloenas nicobarica*, Grey Imperial-pigeon *Ducula pickeringii* and Pied Imperial-pigeon *D. bicolor*, as well as Mantanani Scops-owl. Access is by banca from Rio Tuba but a permit from DENR (Department of the Environment and Natural Resources) at Brooke's Point is required beforehand.

MINDANAO

Mindanao is the second-largest island after Luzon and it lies at the southern end of the archipelago. It is still relatively unknown ornithologically, with many areas not having been visited since the original collectors, and is home to two recently discovered new species, the Bukidnon Woodcock *Scolopax bukidnonensis* and Lina's Sunbird *Aethopyga linaraborae*, and possibly two more. The island was well forested 50 years ago but intensive logging has seriously reduced cover. There are several mountain ranges still with much forest, but lowland forest has been severely depleted. There are also several areas of marsh, the largest being the Agusan and Liguasan Marshes. When visiting the region, it is important to visit both lowland and mountain sites.

Mount Kitanglad Range
This range is best approached from its eastern end in Bukidnon province. The area is large with seven major peaks, and can be accessed from many points. Probably the most convenient is from Dalwangan, 10km north of Malaybalay. At 1,400m is a small lodge where one can stay for a modest fee (please contact the author for details). Practically all Mindanao montane specialities can be found here and it is possibly the best place to see the Philippine Eagle in the wild.

Mount Apo Natural Park
Mount Apo Natural Park, containing the highest peak in the Philippines, is accessible from several points. The better known are at Baracatan, once site of a Philippine Eagle breeding centre, situated at an altitude of 1,000m, and from Kidapawan in Cotabato where PNOC (the national oil company) has developed a geothermal well at 1,500m. The same birds as at Mount Kitanglad can be seen here. Although an Asian National Heritage Park, it is badly deforested.

Paper Industries Corporation of the Philippines
Otherwise known as PICOP, this is a vast logging concession straddling the provinces of Davao del Norte and Surigao del Sur. The area still contains areas of lowland forest plus mid-montane forest up to an altitude of 1,300m. It is rich in birds and is probably the best area for

observing Mindanao lowland forest species. It is excellent for hornbills, pigeons, parrots, pittas, kingfishers and rare flycatchers amongst others. Access and a guide can be arranged either through the author or via the Paper Country Inn at Bislig. The marshes around Bislig airport are good for duck and waterbirds.

Other sites, not so well known, include Mount Busa and the Three King mountains in South Cotabato, east of Lake Sebu, and the Zamboanga Watershed just outside Zamboanga City.

NEGROS

Negros, in the West Visayas, has a main mountain range running down the east side of the island and well-watered slopes on the west. Little forest remains, having mostly been cleared in the 20th century for sugar plantations and later by timber concessions. The little forest that does remain exists at two or three sites and is home to several extremely rare endemics that are found only here or are shared with Panay.

Mount Kanlaon

This mountain, rising to 2,465m, is a national park and lies about an hour's drive east of Bacolod near the resort town of Mambucal. There is a trail that starts just behind the town and climbs up the mountain. Flame-templed Babbler *Stachyris speciosa* and White-winged Cuckoo-shrike *Coracina ostenta* can be seen here; it was also the site of the only known specimen of the Negros Fruit-dove *Ptilinopus arcanus*.

Mount Talinis and Lake Balinsasayao

These cover one of the larger tracts of remaining forest. They lie just north of Dumaguete, the second-largest town in Negros, situated on the south-east coast. Access is by permission from the DENR and the Philippine National Oil Company which has a geothermal well on the slopes of Mount Talinis. The area is the only known site for the Negros Striped-babbler *Stachyris nigrorum*.

CEBU

Cebu is a long narrow island in the Central Visayas with a central mountain range that rises to an altitude of up to 1,000m. Enjoying a good climate, it was settled early by the Spanish, which quickly led to overpopulation and total loss of forest by the end of the 19th century. Recently, one or two tiny remnant patches of forest have been located and some forms of bird once thought extinct have been rediscovered, the most notable being the Cebu Flowerpecker, thought lost since 1906 but rediscovered in 1992.

Tabunan

Tabunan lies in the Cebu Central National Park, in the mountains of central Cebu, and is about an hour and a half's drive from Cebu City. The Cebu Flowerpecker and Cebu Black Shama can be found here.

Olango Island
Olango lies on the eastern side of Mactan Island. Access is by boat from one of several piers. The south-east portion of the island is protected as the Olango Wildlife Sanctuary. The exposed tidal flats and reefs are excellent for egrets and shorebirds such as Chinese Egret *Egretta eulophotes* and Asian Dowitcher *Limnodromus semipalmatus*.

Bohol
Bohol lies to the south-east of Cebu, but in faunal terms it is closely related to Samar, Leyte and Mindanao. Very little forest remains on the island except in the centre, where extensive and successful replanting has led to well-grown secondary forest.

Rajah Sikatuna National Park
This park contains the aforementioned secondary forest and lies alongside the well-known Chocolate Hills. Access is from near Bilar and permission can be obtained from the local DENR office. The geology is basically limestone and the forest is a haunt of species favouring that habitat such as Steere's Pitta *Pitta steerii*. It is also excellent for a large variety of lowland forest species.

PUBLISHER'S NOTE

High-quality photographs of the birds of the Philippines are notoriously difficult to obtain, and many species have never been satisfactorily photographed in the wild state. For the purposes of identification this guide includes several photographs of birds in the hand. These birds were captured by mist netting for ringing and research purposes, and photographed prior to release.

LITTLE GREBE *Tachybaptus ruficollis* 25cm

This is the smallest swimming bird in the Philippines, differing from ducks in its short pointed bill and dumpy tailless body. It swims buoyantly, diving regularly, and inhabits inland marshes, lakes, ponds and other areas of open water where there is sufficient vegetation to hide and nest. The nest is a floating platform made of vegetable matter. The species is not uncommon in suitable habitat. The call, heard at a distance, is a whinnying trill. In breeding plumage the cheek and neck are chestnut with a yellow gape-patch. Otherwise it is brown above and greyish below, more grey-brown in winter.

BROWN BOOBY *Sula leucogaster* 71cm

This large gannet has a long cigar-shaped body, strong pointed bill and wedge-shaped tail. It flies strongly with alternate flaps and glides, often climbing high before plunging at an angle into the sea for fish. The adult is unmistakable, being all brown except for a white belly and bar on the underwing. The species is not uncommon in the Sulu Sea, the Pacific and off the south coast of Mindanao; it is usually seen out at sea but can appear offshore after storms. It nests in colonies on the ground on small secluded islands.

ORIENTAL DARTER *Anhinga melanogaster* **89cm**

Looking like a streamlined cormorant with its long thin neck and long fan-shaped tail, the Darter inhabits extensive areas of water with large fringing trees in which to perch with wings held out to dry. It usually breeds colonially, placing its stick nests in the tops of small trees near water. It pursues fish under water, often spearing them with its sharp bill. It has a brown head and neck with a white streak down the side, but is otherwise black with silver-grey streaks on the mantle. Although widespread in South-East Asia, it has become very rare in the Philippines.

LESSER FRIGATEBIRD *Fregata ariel* 76cm

This large aerial seabird spends much of its life on the wing in the open ocean, flying high and diving after other seabirds to rob them of their prey. The silhouette is distinctive with long narrow wings and deeply forked tail. In identifying frigatebirds, care is needed to determine the amount of white underneath. The male is all black with a red gular patch and white flanks and wing-base. The female has a white breast and concave belly-patch, some white at the base of the underwing and a black chin. The birds roost on trees and fishtrap platforms and nest in colonies on small islands.

GREAT-BILLED HERON *Ardea sumatrana* 117cm

This very large heron is widespread from South-East Asia and the Philippines south to Australia, but is nowhere common. Usually found singly or in pairs in remote coastal regions, breeding in mangroves and feeding on exposed reefs and mudflats, it is elusive and has a large home range. It is grey-brown all over with grey feet and a large heavy black bill with which it stabs its coral fish and crustacean prey. The call is a loud croak or a repeated roar. An excellent place to see this bird is Rasa Island, just off Palawan.

PURPLE HERON *Ardea purpurea* 92cm

This fairly large and not uncommon heron is found throughout the country in marshy areas, mangrove swamps and dry grasslands where it occurs singly, although it nests in colonies on the ground in dense cover such as reedbeds. It is not easy to see as it stands motionless with neck stretched, looking like a clump of vegetation. It occasionally perches atop dead trees. Basically crepuscular, it feeds in the evening and early morning. It is grey on the back and wing-coverts, with black flight feathers; otherwise it is reddish-brown with a diagnostic black line down the neck. It flies slowly and heavily with head held back.

GREAT EGRET *Egretta alba* 90cm

At nearly one metre tall this is the largest egret, its size noticeable in flight, with bill and distinctly kinked neck 50% longer than its body. In winter the bill is yellow, with green facial skin and black legs; when breeding, the bill turns black, the face bluish and thighs greenish. Widespread from Europe to Japan, it was traditionally a winter migrant to all parts of the Philippines, appearing alongside other egrets in open wetlands, marshes, coastal mangroves and flooded paddies. Recently, flocks have been seen in summer in breeding plumage, indicating that it may breed.

LITTLE EGRET *Egretta garzetta* 65cm

A common winter visitor from China, Korea and Japan, this elegant shorebird is found throughout the Philippines from late September to April in various habitats including tidal mudflats, fishponds, inland marshes and flooded paddies where it feeds on fish, crustacea and other aquatic animals. All white, it can be identified by a combination of size, long black bill and legs with yellow feet, although the latter may be hard to see if covered in mud. From February onwards, it starts to grow its two long plumes on the nape as part of its breeding dress.

CATTLE EGRET *Bubulcus ibis* 48cm

Found in wetland areas, marshes, rice paddies and agricultural land with scattered trees to roost in, this bird is usually seen near animals, particularly cattle and water buffalo on which it perches. Resident throughout the Philippines, it is joined for the winter by huge flocks from China in October and November. Its nest is well hidden in a large clump of tall grass. Stocky with a thick neck and heavy jowl, it is all white in winter with a yellow bill and dusky legs. When breeding, it has buff-coloured tufts on the head, neck and lower breast, and orange-buff plumes on the back.

CHINESE EGRET *Egretta eulophotes* 70cm

The total population of this rare egret is only around 2,500 birds. The Chinese Egret breeds in coastal China and Korea and winters mostly in the Philippines. It inhabits areas which combine expansive intertidal mudflats, where it feeds, and mangroves, where it roosts in dense groups. When feeding, it is very active, running after its prey with wings open, unlike the similar Eastern Reef Egret, which is sluggish. In winter it has greenish legs, bill and facial skin. In breeding plumage, from March onwards, it sports a long tuft of white feathers on the nape, the legs are black with yellow feet, and the bill is yellow with bluish facial skin.

PACIFIC REEF EGRET *Egretta sacra* 58cm

This solitary bird can be found on rocky shorelines and exposed reefs. It stands stock-still, crouched forward on short legs in search of fish and crustaceans. Not uncommon in suitable localities, it is often seen flying low over the water to another feeding site. The nest is made of sticks set amongst the rocks near the tideline in an undisturbed area. It occurs in two morphs, one all white and one uniform slate-grey with a white chin; the former is rare in the Philippines. The legs are green and the bill black, turning to yellowish-orange when breeding.

LITTLE HERON *Butorides striatus* 45cm

This small heron is fairly common on coastal reefs and mudflats, in mangroves and marshlands. Other, larger members of the species inhabit forest streams and are probably northern migrants. The birds sit quietly on exposed rocks or stumps during the day and, when disturbed, fly low to a hiding place, giving a single loud 'kwerk' in flight. They feed by night on crustaceans, worms and fish. They are dark grey above, paler and buffier below, with a greenish gloss and buffy edges to the wings, and a long floppy greenish-black crest. The long bill is greenish-yellow, the legs green. White ear-coverts and black malar stripe are also diagnostic.

RUFOUS NIGHT-HERON *Nycticorax caledonicus* 63cm

Although not uncommon, this heron is crepuscular and rarely seen. In the evening it flies, giving a raucous call, from its roosts in the mangroves or nearby forest to areas such as rice paddies, drained fishponds or shallow riverbeds, where it feeds on frogs, fish and crustaceans. It breeds in colonies in tall trees often with Purple Herons, egrets and, more recently, Black-crowned Night-herons *Nycticorax nycticorax*. It has a black and yellow bill, black crown, chestnut back, rufous rump, wings and tail, cinnamon throat and breast, white belly and greenish legs. Young birds are brown, heavily streaked and spotted.

YELLOW BITTERN *Ixobrychus sinensis* 38cm

This small bittern visits the Philippines from October to April, returning in spring to the Asian mainland to breed. It is quite common in suitable habitat, which is inland wetlands and marshes, but it hides in the long grass and reeds, often perched motionless on the stems although quite active during the day. It has a distinctive flight pattern, the black flight feathers contrasting with fawn wing-coverts and dark brown mantle. Adults have black caps and tails, light brown upperparts and buffy underparts, whilst juveniles are heavily streaked above and below.

CINNAMON BITTERN *Ixobrychus cinnamomeus* 39cm

Resident and fairly common, the favoured habitat is large dense clumps of grass in and around marshlands, fishponds, riverbanks and even dense-grass areas away from water. Crepuscular, it perches motionless with its neck straight up to escape detection. In flight, it can be told by its uniform cinnamon wings and tail. Males are a bright cinnamon on the upperparts, buffy-orange on the underparts, with a central line of black streaks on the breast and dark pectoral tufts. Females are darker and browner with more streaking below and a sooty cap. Young are like females but have yellowish-buff spots above and are more streaked.

SCHRENCK'S BITTERN *Ixobrychus eurythmus* 34cm

This bittern is a scarce but secretive winter visitor to the Philippines from north-east Asia and Japan where it breeds in extensive reedbeds. It appears to winter in forests, emerging at night to feed in marshes, paddies and lakesides. The male is separated from the Yellow Bittern *Ixobrychus sinensis* by its dark brown (almost black) back, dark stripe from throat to belly, and less contrast between its buffy wing-coverts and brownish-grey flight feathers. The underwing is pearl-grey, the tail brown and crown and nape blackish-brown. The female is rufous-brown with white flecking above, buff-streaked dark-brown below.

WOOLLY-NECKED STORK *Ciconia episcopus* 75–91cm

This large stork, which ranges from India to Sulawesi, is the only one to breed in the Philippines, where it is critically endangered, having once been widely distributed. In the past twenty years it has only been recorded from eastern Mindanao and north-east Luzon, although it may occur in some other remote areas. It is usually seen soaring high overhead in lowland forest with open areas or feeding in ricefields, marshes and gallery forests on reptiles, frogs and fish. The entire upperparts and breast are black glossed with green and purple, and the fluffy neck, lower belly and long undertail-coverts are white.

GARGANEY *Anas querquedula* 35cm

The Garganey is a small dabbling duck that breeds in north Asia and winters in the tropics including the Philippines, arriving late October/early November at such places as the Candaba Marshes where it can be seen in tens of thousands before dispersing throughout the country. It prefers expanses of fresh water with dense vegetation where it can feed and hide. The males can be identified by their broad white head-stripe and, in flight, their blue-grey forewings. The females are mottled brown with a distinct head pattern consisting of a dark eye-stripe, obvious pale supercilium and white throat.

NORTHERN SHOVELER *Anas clypeata* 50cm

The Shoveler breeds in the northern hemisphere from Europe across to north-east Asia and China. Part of the population winters in the Philippines, arriving in fair numbers around mid-October. It favours freshwater lakes, ponds and flooded wetlands but can also be seen in sheltered bays and undisturbed rivers and often associates with other wintering duck. Its spatulate bill, which is longer than the head and visible from afar, is diagnostic; it is used as a sieve as the bird swims forward. The male is unmistakable with its greenish-black head, white breast and chestnut belly, whilst the female is mottled brown with a dark line through its eye.

TUFTED DUCK *Aythya fuligula* 47cm

The Tufted Duck is a winter migrant to the Philippines from the north, arriving in late November and departing in April. It can be found throughout the country in small numbers on undisturbed lakes, reservoirs and pools in marshy areas, swimming buoyantly and diving frequently for its food. It has a distinct square rounded head and broad bill. The male is easily recognised by its generally black plumage with white flanks, whilst the female is mainly brown with a small amount of white at the base of the bill. Both sexes in winter have a short crest, whilst in flight they are easily told by their white bellies and white band across the flight feathers.

WANDERING WHISTLING-DUCK *Dendrocygna arcuata* 49cm

Widespread from Borneo to Australia, it is also found throughout the Philippines in well-vegetated, flooded marshy areas, lakes and ponds where it can swim and hide, but also selecting areas where it can perch on stumps or dead matter. It is very noisy, giving twittering whistles as it rises in small flocks, circling with drooping narrow necks before it drops back into cover. It is blackish-brown above with a white band on the rump and undertail-coverts. The underparts are paler and more rufous. The underwing is dark, not pale as in the Philippine Duck. The grass nest is placed in tussocks of marsh reeds growing on small hummocks.

PHILIPPINE DUCK *Anas luzonica* 51cm

This is the only endemic duck in the Philippines, found throughout in large tracts of well-vegetated marshlands, flooded paddies and undisturbed slow-moving rivers, where it feeds usually at night. Although mostly in small numbers, large flocks congregate in sheltered bays to roost during the day, flying inland to feed at night. The nests are made of grass and placed in dense clumps near water. It is a fairly large, heavy duck, grey-brown except for the rufous face, neck and throat, and dark brown cap and eye-stripe. In flight, the whitish underwing lining and dark flight feathers are diagnostic. The call, probably given by the female, is a 'quack-quack'.

OSPREY *Pandion haliaetus* 55–61cm

This, the sole member of a worldwide family, occurs throughout the Philippines. It is a fish-eating raptor and thus usually found near water, both inland and coastal. It often hovers before diving at an angle to hit the water feet-first, and is often seen carrying its prey in its talons back to a favourite perch such as a tall dead tree or post in the sea. One race is resident and another, brown-crowned, is a migrant from north Asia. The upperparts are blackish-brown with blacker wings, and the head is white with a black band through the eye to the nape. The underparts are white with a dark brown breast-band. In flight, the long narrow kinked wings are distinctive.

BRAHMINY KITE *Haliastur indus* 43–51cm

This striking Asian raptor occurs throughout the islands in open country near water, and often near towns, where it can be seen either perched on a tall tree or soaring high in the air. It has been heavily persecuted especially for the pet trade. It eats fish, lizards, crustaceans, small mammals, carrion and offal. Adults are unmistakable: the head, neck and underparts are white with black shaft-streaks, whilst the rest of the body is bright chestnut; the wing-tips are black. Young birds are brown with a buffy-white head.

WHITE-BELLIED SEA-EAGLE
Haliaeetus leucogaster 70–76cm

This large thick-set Asian eagle is an uncommon but widespread resident in coastal areas and up large rivers, where it catches fish and sea-snakes. The nest, often re-used over many years, is a huge structure of sticks built at the top of a tall dead tree, rocky outcrop or even electricity pylon. The bird is unmistakable as it soars on broad wings held forward and in a shallow 'V'; the wedge-shaped tail is also distinctive. The adult has a white head, underparts, underwing lining and tip of tail, black flight feathers and base of tail, and brownish-grey tail. The young is dark brown with buffy head and black flight feathers. The call is a clanking, repeated 'honk'.

PHILIPPINE SERPENT-EAGLE *Spilornis holospilus* 51cm

Found throughout the country with the exception of Palawan, where it is replaced by Crested Serpent-eagle *S. cheela*, a widespread southern Asian species. Both are found around forests and plantations at all altitudes and eat mainly snakes and other reptiles. Both have distinctive, loud, far-carrying whistled calls, in the former a lazy 'seee-ap weep weep', louder and shriller at the start, in the latter 'eep-a-pa weep weep weep', with the first note quieter. Similar in appearance, they have a distinctive white subterminal band on the underwing, while at rest the broad crest, yellow cere, fine white spots on the wing-coverts and underparts, and the unfeathered tarsi are good marks.

PHILIPPINE EAGLE
Pithecophaga jefferyi 99cm

This huge eagle, equal largest in the world, is endemic to Luzon, Mindanao, Samar and Leyte. Owing to loss of habitat and hunting, it may be down to a few hundred birds. It is found in forested mountain valleys, gliding from tree to tree, seeking prey such as flying lemurs, monkeys, squirrels and other smaller animals. It nests high up in the forks of large trees, and lays one egg every two years, often around October to November. The upperparts are a rich brown with pale margins to the feathers, and the underparts are buffy-white. It has a highly arched, powerful bill and a distinctive yellowish crest.

RUFOUS-BELLIED EAGLE *Hieraaetus kienerii* 46cm

This medium-sized eagle is found from India through South-East Asia to the Moluccas, and is thinly distributed throughout the lowland and mountain forests of the larger Philippine islands. It often soars over the forest canopy and occasionally dives with wings closed at high speed. When at rest, it favours branches of tall trees just below the canopy on a forested slope. In flight, the white throat and upper breast contrast with the black head and rufous belly. The immature is white below with a black patch on the flanks and a distinctive black mask. The call is a series of five or so rising notes, often preceded by a scream.

PHILIPPINE FALCONET *Microhierax erythrogenys* 16cm

This tiny falcon is fairly common on Luzon, Mindanao, Mindoro and the Visayan Islands but is absent elsewhere. It inhabits the clearings and edge of lowland primary and secondary forest. Seen singly or in pairs, it perches on prominent branches on old dead trees from which it hawks for flying insects. It nests in old woodpecker holes. The upperparts, flanks and thighs are a glossy blue-black, the underparts white, the bill and feet black. The call is a noisy 'kek-kek-kek-kek' which it gives as it dives for prey.

PHILIPPINE HAWK-EAGLE *Spizaetus philippensis* 61cm

Found throughout the Philippines except Palawan, this raptor inhabits lowland and mountain forests, where it soars over the canopy, announcing its presence with a distinctive two-note whistle, 'wheet-whit'. When perched, it displays a long black crest. The pale buffy feathers on the head and neck have dark brown centres giving a streaked appearance, and the tail is light brown with dark brown bars. The rest of the upperparts are dark brown. The rich brown thighs with fine white bars are diagnostic. The white throat is washed rufous and streaked brown; in the larger Luzon race, the lower breast and belly are brown, whilst southern birds have the belly barred black, white and brown. In adults the underwing is finely barred black-and-white whilst in young birds it is all white.

TABON SCRUBFOWL *Megapodius cumingii* 35cm

The chicken-like ground-living tabon birds, as they are locally known, use their large powerful feet for piling up mounds of vegetation in which they lay their eggs, allowing the heat of the decaying matter to do the incubation. Found on Sulawesi, islands off Borneo, and the Philippines, this much-hunted gamebird occupies small islands and coastal forest on the larger islands. At dawn and dusk it gives a loud eerie whistle like a works siren, but is very secretive, running away at the first sign of danger. The upperparts are rich olive-brown, the underparts dusky blue-grey. The head and neck are grey with a pink face and throat.

BLUE-BREASTED QUAIL *Coturnix chinensis* 13–15cm

This very small plump quail is widespread from India to Australia. In the Philippines it is common in open grassland, fallow paddyfields and similar habitat mainly in the lowlands. It is difficult to spot; moreover, it is partly nocturnal. When flushed, it flies off for about 30m or so before alighting, the male appearing quite black. Males are blackish-brown above with white shaft-streaks; below they show a distinct black bib with white marks, the sides of head, breast and flanks are deep blue, the belly and undertail chestnut. The female is paler, streaked brown above, buffy below with black bars on breast and flanks and a buffy eyebrow.

RED JUNGLEFOWL
Gallus gallus M 66cm, F 42cm

This bird is found in lowland and mountain forests throughout the Philippines, feeding on the ground on grains, seeds, insects and worms. Usually solitary, it roosts with others high up in a tree. The nest is a hollow in the ground placed under a bush. The male is very like the native domestic chicken but is slimmer, slightly smaller, and normally with green-grey legs; it has a whitish 'puff' at the base of the tail and white ear-patches. The female is brown, mottled with rufous and ochre-yellow streaks on the neck. The call 'ka ka-kaa ka' recalls the domestic fowl but is shorter and sharper, the last note not so stretched out.

PALAWAN PEACOCK-PHEASANT *Polyplectron emphanum* 45–51cm

This splendid bird is restricted to lowland primary forest on the island of Palawan where it is very secretive and difficult to see, feeding on a variety of fruits, seeds, insects and other small animals. The typical call is a loud, harsh, raspy single note repeated every few seconds. The male has a long blue crest, white ear-patches and eyebrow, and black head, underparts and flight feathers. The mantle and wing-coverts are metallic blue; the back, rump and tail are black with fine gold specks, the tail with two rows of blue ocelli bordered black and grey. The female is brown but retains the blue ocelli on the tail.

SPOTTED BUTTONQUAIL *Turnix ocellata* 18cm

Large for a buttonquail, this bird is endemic to Luzon, where it is relatively common, with one record from Negros. It favours open scrub near forests or bamboo groves; it feeds on the floor of dry open forest but also occurs in areas with scattered trees and bushes. The blackish head has three white stripes, the upperparts are brown mottled grey and black, the wing-coverts are buffy with large black spots and the rest of the wing is grey-brown; the breast is rufous, the belly brown. The male has a white throat and sides of head; these are black in the female, which also has a broad chestnut collar.

SLATY-LEGGED CRAKE *Rallina eurizonoides* 27cm

Although not uncommon and found throughout the country, this rail is very difficult to see, being fairly nocturnal and, in contrast to other family members, found foraging on the floor of forest and dense scrub with patches of trees. Its presence is usually revealed by its strange frog-like nasal call, a single 'onk' repeated constantly, often in duet with another bird. It usually creeps away at the first sign of danger, but will fly up into a tree if surprised. It is earthy brown above, barred black-and-white below, with a white chin and rufous head, neck and throat.

WHITE-BROWED CRAKE
Porzana cinerea 21cm

This small Asian-Pacific crake is common and widespread in the Philippines in marshlands, overgrown paddies and the vegetated areas around lakes and ponds. It walks nimbly over floating vegetation and flies to cover when flushed. It is quite noisy, several birds suddenly calling from within deep cover, a high-pitched piping 'cutchi-cutchi-cutchi'. Brown above and white below with buffy flanks and undertail-coverts, it has a distinct head pattern consisting of a dark grey crown, black line through the eye with a short white eyebrow above and white line below. The short bill is yellow and legs greenish.

WHITE-BREASTED WATERHEN *Amaurornis phoenicurus* 28cm

Although not often seen, this rail is common throughout the Philippines, its distinct call a loud monotonous 'wok wok wok wok' or 'ker-wak-wak', often being heard even close to built-up areas. It is less shy in Palawan, where it can be seen walking around on roads during the day. It inhabits open country, rough pastureland and marshes, preferring wetter areas than the Plain Bushhen *Amaurornis olivacea*. The adult bird is slaty-black above with a white forehead, face, throat and centre of belly and light rufous flanks and undertail-coverts. Immatures are browner and the white underparts dirtier.

PHILIPPINE SWAMPHEN *Porphyrio pulverulentus* **40cm**

This large gallinule is found on Luzon, Mindanao and several other larger islands in well-vegetated reed-lined swamps and wetlands, walking over the floating plants, clambering amongst the reedbeds, flicking its tail to reveal its white undertail-coverts. Looking like an outsize Common Moorhen without the white band, it has huge red legs and red bill and frontal shield. Generally bluish-black in plumage, paler around the head, it is rich olive-brown on the back, rump and scapulars and blackish-green on the flight feathers. It is not uncommon in suitable habitat such as the marshes fringing Laguna de Bay, but owing to its size it is extensively hunted for food and consequently much rarer than before.

COMMON MOORHEN *Gallinula chloropus* 36cm

The white streak across the top of the flanks and white outer undertail-coverts are diagnostic of this largely black, medium-sized gallinule, which also has a red frontal shield and bill with a yellow tip. Young birds are olive-brown, paler and mottled on the underparts, notably the throat. It is common throughout the Philippines, in a variety of wetland areas where it walks stealthily around on floating matted vegetation flicking its tail. The call is a disyllabic 'kurruk' or 'kittik'. The nest is usually made of sticks and reeds, well concealed close to the water's edge.

PACIFIC GOLDEN PLOVER *Pluvialis fulva* 23cm

Breeding in Siberia and western Alaska, this common winter visitor to the Philippines is found, sometimes singly, often in small flocks, in habitats such as coastal mudflats, exposed reefs, mangroves, inland marshes, paddyfields and short grassland. In winter the upperparts are dark buffy-brown, appearing uniform in flight, with a buffy forehead, eyebrow, sides of head, neck and breast. The centre of throat to belly is buffy-white. In breeding plumage the bird is black below, black and gold-spotted above, with a white eyebrow running into a broad band down the side of the neck. The call is a clear 'too-weet'.

LITTLE RINGED PLOVER *Charadrius dubius* 17cm

This neat shorebird is told from similar plovers in flight by its call, a clear far-carrying 'peeyoo', lack of wing-bar and, at rest, from Kentish *Charadrius alexandrinus* and Malaysian Plovers by its full black breast-band and hindneck, and yellow legs and eyering. When breeding it shows a black mask with white forecrown and short line above the eye; the upperparts are greyish-brown. In winter the black is replaced by brown. Both resident and visitor, it is found commonly throughout the islands in inland marshes, paddyfields, stony riverbeds and on the coast, breeding on damp stony ground, the nest being a hollow in the ground.

MALAYSIAN PLOVER *Charadrius peronii* 16cm

One of the few resident shorebirds of the Malay Peninsula, Sundas, Sulawesi and the Philippines, it is uncommon and local, occurring in pairs on secluded coral-sand beaches and easily disturbed by human encroachment. Its blotched creamy-buff eggs are laid in a scrape in the sand. The male is sandy-brown above, white below, with a white collar over a black band that extends from breast-side to hindneck, a black spot behind the eye and a black mark on the head; in flight it shows a white wing-bar. The female is similar but has rufous instead of black. The legs are grey. The call is a soft 'twik'.

GREATER SAND PLOVER *Charadrius leschenaulti* 21cm

This small plover breeds in Central Asia and is a common passage migrant and winter visitor to coastal mudflats, sandbars and drained fishponds. It usually occurs in small flocks with other plovers such as the very similar Lesser Sand Plover *Charadrius mongolus*, from which it differs in winter by its larger size, darker legs and longer, thicker black bill, and in spring by its narrower and fainter rufous breast-band with no black dividing throat-line. In winter it lacks the black head and neck markings and white hindneck of other small plovers. In flight, it shows a narrow white wing-bar; its call is a soft, trilling 'chirrirrip'.

WHIMBREL *Numenius phaeopus* 44cm

A winter visitor from northern Siberia, this large shorebird occurs throughout the coastal Philippines from late September through to April, usually on mudflats, exposed reefs and mangrove areas, but also sandbars and rocky shores. It is mottled brown above with a long decurved bill, a distinct head pattern consisting of a buffy crown-stripe and eyebrow and dark eyestripe, and white lower back and rump. The underparts are mostly white but buffy barred brown on flanks and underwing, and the sides of the face and neck and breast are buffy streaked brown. It gives a clear rapid series of whistles, 'titititititititi'.

BAR-TAILED GODWIT *Limosa lapponica* 38cm

Godwits are large waders with long straight bills for extracting worms in mudflats. Bar-tails differ from Black-tails *Limosa limosa* by their smaller size, lack of wing-bar in flight, and shorter, upturned, two-toned bill; the Asian Dowitcher *Limnodromus semipalmatus* is smaller, with a black bulbous-ended bill. The Bar-tail is an uncommon visitor to the Philippines from its Arctic breeding grounds, but it is regular on Olango Island. It has a white back and rump, and is brick-red below when breeding; in winter it is mottled brown above and whitish below with a grey-washed breast. The call is an excited 'kew-kew'.

COMMON GREENSHANK *Tringa nebularia* 32cm

The first sign of the Greenshank is its far-carrying strident three-note whistle, 'tew-tew-tew'. It is a nervous, medium-large wader with a fairly long, slightly upturned bill. In winter it is pale grey above, white below, with a distinct white wedge up the back which contrasts with the unmarked wings in flight. The legs are green. It occurs singly or in small groups, often associating with other waders, and is found not only on coastal mudflats, estuaries and fishponds but also inland on rice paddies, riverbanks and elsewhere. The Marsh Sandpiper *Tringa stagnatilis* is very similar but smaller and more delicate, with a black needle-like bill.

COMMON REDSHANK *Tringa totanus* 28cm

A passage migrant and winter visitor from northern Asia, where it breeds in meadows, moors and saltings, the Redshank is found fairly commonly throughout the country either singly or in small groups of four or five, in coastal areas such as exposed mudflats, fishponds and estuaries. The red or orange legs and base of bill distinguish it from all waders except the Spotted Redshank. The tip of the bill is black, and in immature birds the legs are yellower. In winter, it has grey-brown upperparts and breast, the latter with brown streaks. The breeding plumage is similar but the upperparts are heavily streaked. In flight it shows a broad white trailing edge on the inner wing, a white rump that extends as a wedge up the back, and a finely barred black-and-white tail. It is active and noisy, bobbing up and down when disturbed and then springing up in the air with a piercing three-note call, 'tleu-hu-hu', as it flies off.

WOOD SANDPIPER *Tringa glareola* 21cm

This medium-sized sandpiper is one of the commonest waders wintering in the Philippines, occurring in loose flocks throughout the country's inland wetlands and wet paddies between October and April. A nervous bird, it flushes easily, towering into the air calling shrilly 'wi-wi-wi' before landing 50m or so on. It has brown upperparts with pale speckles. To separate it from the darker-backed Green Sandpiper *Tringa ochropus*, note the pale supercilium and eyering, less contrast between the white uppertail-coverts and brown back, and pale underwing. The rump is dark and there is no wing-bar.

BUKIDNON WOODCOCK *Scolopax bukidnonensis* 33cm

Found in montane forest above 1,000m on Luzon and Mindanao. Shy and secretive, it rests by day on the forest floor and is active at night. From January–March, it engages in a distinctive aerial 'roding' display at dusk and just before dawn. Birds fly in a wide circuit over the forest, giving a loud, metallic, rattling 'pip-pip-pip-pip-pip' interspersed by very quiet grunts. Rich reddish-brown above, finely barred and vermiculated with black and broadly barred blackish across the crown, the underparts are paler and buffer. The eye is placed high and far back on the head, and the bill is long with a flexible tip to extract worms and other invertebrates from the soil.

COMMON SANDPIPER *Actitis hypoleucos* 20cm

Common is a good word for this winter visitor from the Palearctic, found throughout the islands in a variety of habitats including coastal shores, exposed reefs, rivers, rice fields and reservoirs. It is easily identified by its broad white wing-bar and distinctive flight in which the wings are flickered downwards between glides. The call is a plaintive piping 'twee wee wee wee'. When feeding, it constantly bobs its tail. The upperparts, rump and central tail are uniform olive-brown; the sides of the tail are white. The underparts are white except for the sides of the breast and neck, which are brown.

GREY-TAILED TATTLER *Heteroscelus brevipes* 25cm

This medium-sized wader is a common winter visitor from its north-east Siberian breeding grounds. It arrives around September, staying until May, and can be found throughout the country along the coast on mudflats, reefs, mangrove channels and rocky shores either singly or in small groups, often with other waders. In winter it is uniform grey above, white below but slightly washed grey on the breast. The yellow legs, white supercilium and unmarked wings are diagnostic. In breeding plumage the underparts are barred except the lower belly and undertail-coverts. It flies with clipped wing-beats and a strong double whistle 'tooo-eeet'.

RUDDY TURNSTONE *Arenaria interpres* 21cm

This stocky medium-small wader with orange legs and variegated plumage is quite common in winter, on mudflats, exposed reefs and rocky shores, turning over pebbles and weeds to expose small crustaceans, crabs and other food. In breeding plumage it has a distinct harlequin appearance with a black and white head, black throat and breast, and black and tortoiseshell upperparts. In winter it becomes a mottled grey-brown. Throughout the year it keeps its pied upperparts with the black and white lower back, rump and tail, white wedges on the lesser wing-coverts and white wing-bar. It often associates with other waders; the call is a sharp 'kititit' or 'keeu'.

LONG-TOED STINT *Calidris subminuta* 16cm

One of the smallest waders in the Philippines, this is a fairly common migrant from breeding grounds in east Siberia. It is rarely seen on the coast but favours inland marshy areas, wet paddyfields and drained fishponds where it forages on the exposed mud in small groups with other waders. It resembles a small Sharp-tailed Sandpiper *Calidris acuminatus*. In winter it is told by the bold black streaks on the brown upperparts, finely streaked buffy-grey neck and breast, white underparts and yellow or green legs; in summer the upperparts and breast are tinged rufous. The streaked crown and pale supercilium give it a capped appearance. The call is a purring 'chrrrup'.

GREATER PAINTED SNIPE *Rostratula benghalensis* 25cm

This colourful wader is unusual because the female has the brighter plumage and does the courting, while the male incubates the eggs. It ranges from Africa to Australia and occurs throughout the Philippines in swampy ground and marshes. The nest is a hollow on dry land close to the water. It feeds mostly at night and skulks by day. The female has a dark chestnut head and breast with a white eye-patch and yellow crown-stripe. Her territorial call is like the sound made by blowing into an empty bottle. A bold white 'V' extends from the back over the shoulders. The male is smaller and duller, more mottled with golden spots on the wings and a yellow eye-patch.

BEACH THICK-KNEE *Esacus magnirostris* 51cm

This large plump shorebird occurs from Malaysia and the Philippines to Australia. Now very rare, it is secretive and nocturnal, inhabiting remote sandy beaches and small island reefs. Its call, usually given at night, is a mournful 'wee-loo', and it has a curious head-bobbing movement. It uses its powerful black yellow-based bill for feeding on sand crabs. Crown and upperparts are greyish-brown, side of head boldly marked with black and white stripes. The grey upperwing contrasts with the white inner and black outer primaries. Underparts and underwing are white, legs, feet and eye yellow.

ORIENTAL PRATINCOLE *Glareola maldivarum* 23cm

Usually seen over open marshy country hawking for insects in a graceful tern-like flight, this bird shows a long forked black-tipped tail and white uppertail-coverts and rump. At rest it is identified by its rich brown plumage, creamy throat bordered with black, and short legs, but it is easily lost to view when it settles on bare earth. It breeds in small numbers in the Philippines, around April and May, its nest being a small indentation in a field. The population swells in the autumn with incoming migrants, and in October and November huge numbers sometimes form compact spirals as if preparing to fly south.

BLACK-WINGED STILT *Himantopus himantopus* 36cm

Two populations of Black-winged Stilt commonly visit the country, one from the north with a blackish head and one from the south with a strongly marked black hindneck; this latter may breed in Mindanao. Both forms occur in brackish and freshwater swamps, flooded paddies, fishponds and coastal flats, and are usually in flocks which can be quite large. Very skittish, they fly off noisily, giving a loud high-pitched yelping call, 'kik-kik-kik'. Their pink legs are the longest, proportionately, of any bird in the Philippines, giving them a high graceful carriage and being very noticeable in flight.

BLACK-HEADED GULL *Larus ridibundus* 38cm

This is an uncommon migrant to large estuaries, mudflats and occasionally inland lakes and marshes on some of the larger Philippine islands. It is normally found in loose groups. The back and wings are light grey except the outer primaries, which are white with black tips; the wing is grey below with black-edged primaries. The rump, tail and underparts are white. The head is white with a black mark behind the eye in winter, but when breeding it turns chocolate-brown with a broken white eye-ring. Young birds are like non-breeders with brown bars on the inner wing and a black band on the tail.

GULL-BILLED TERN *Gelochelidon nilotica* 38cm

This medium-large tern is told by its pale upperparts, stout black bill and slightly forked tail. In winter it is white below and grey above with grey mottling on the nape and a black patch through the eye. The entire cap is black when breeding. It hawks and skims fairly low over the water, rarely plunging. It has an almost worldwide range, breeding in the north and wintering in warmer waters. In the Philippines it is uncommon and local, usually on the coast in sandy estuaries and tidal flats, but sometimes in wet areas inland. Olango Island is a regular haunt.

BLACK-NAPED TERN *Sterna sumatrana* 33cm

Terns are sea-going birds with long pointed wings, forked tails, fine pointed bills and a graceful, buoyant flight. This species is gregarious, often in tight flocks, found both near the shore and out to sea and feeding by diving or picking prey off the surface. Not uncommon in favoured areas, it breeds on offshore rocks and small islets, the nest being a scrape in the sand or shingle. It is a fairly small, white tern, slightly grey above, with a very long pointed tail and a distinctive black nape-band and narrow yellow-tipped black bill.

BROWN NODDY *Anous stolidus* 38cm

The graceful sea-going noddies differ from other terns by being a uniform sooty-brown. They are fairly heavy-bodied and broad-winged and have a wedge-shaped, notched tail. The crown and forehead are light grey which blends into grey-brown on the hindneck. The rarer Black Noddy *Anous minutus* is smaller and blacker than the Brown, and the grey crown is paler. Usually seen in flocks out to sea, their flight is lazy and wheeling, dropping down to pluck small fry from the surface. Locally common, they breed on offshore islands such as Tubbataha Reef in the Sulu Sea.

WHISKERED TERN *Chlidonias hybridus* 25cm

This marsh tern is common in the Philippines between October and April when it departs north to breed, although some Australian breeders summer here. It is seen inland over ponds, rice fields, marshes and sometimes coastal mudflats, catching prey by shallow plunges or low skimming flights. The breeding adult has a black crown and forehead and grey underparts contrasting with a white cheek-patch. Non-breeding adults have white underparts and forehead, and grey wings, back and uppertail-coverts. The species differs from the White-winged Black Tern by its grey rump, lack of black cheeks, and blacker crown.

WHITE-WINGED BLACK TERN *Chlidonias leucopterus* 23cm

This small marsh tern breeds from Europe to Siberia and winters from Africa to Australia. In the Philippines it appears to be more of a passage migrant, although some birds may winter in inland marshes, wet paddyfields and fishponds. In May flocks pass through already in their breeding plumage – black head, body and wing linings, grey wings and white tail, rump and vent. In winter the species is grey above and white below, and told from non-breeding Whiskered Terns by its more complete white hind-collar, less black on crown and separate black ear-coverts. It feeds by fluttering over the surface and dipping its bill in the water.

POMPADOUR GREEN-PIGEON *Treron pompadora* 28cm

This medium-sized pigeon occurs throughout the Philippines except Palawan in lowland to mid-altitude forest, usually at the edges. It is particularly active in the early morning, flying fast and direct in small groups to feed in fruiting trees. In flight it appears short and stocky with pointed wings. It roosts in large numbers in tall trees where it is well camouflaged. The call is a variety of modulated mellow whistling notes. Basically green in colour, it has a bright maroon mantle and black wings with two yellow bars. The female lacks the red mantle. Yellowish-white undertail-coverts distinguish it from the Thick-billed Pigeon *Treron curvirostra*.

PINK-NECKED GREEN-PIGEON *Treron vernans* 25cm

Widespread and fairly common, small flocks are found in coastal scrub, mangroves and more cultivated areas with adjacent dense scrub, rarely above 300m. Birds perch close together atop tall bushes and can be picked out when one scans the area. Nests are loosely built of twigs 2–10m from the ground. Smaller than the Pompadour Green-pigeon, the male has a blue-grey head, grey mantle, green back, chestnut uppertail-coverts and grey tail, whilst below it has a pink neck and upper breast, orange lower breast, yellowish-green belly and chocolate undertail-coverts. The female has a blue-grey and dark tail with buffy or pale cinnamon undertail-coverts.

WHITE-EARED BROWN-DOVE
Phapitreron leucotis 23cm

This small, common and rather tame pigeon is a member of a distinct genus of brown fruit doves endemic to the Philippines. It inhabits forest, plantations and more open country with scattered trees. Although it does not form flocks, many birds can be seen together in fruiting bushes often near the ground. It is basically brown with bronze reflections. The crown is grey, as are the undertail-coverts and tip of tail. A black line passes from gape to nape under the eye, and below this is a white line. The call is a series of descending and accelerating hoots, easily imitated by small boys who capture the birds as pets.

AMETHYST BROWN-DOVE *Phapitreron amethystema* 27cm

This medium-large pigeon occurs on most larger islands with the exception of Palawan but is generally rather scarce on Luzon and Mindanao and even rarer elsewhere. A forest bird, it is found in the lowlands but probably prefers middle and high elevations at 500–2,000m. It is most often seen singly or in pairs, in and around fruiting trees. The call is a deep, sonorous 'hoop' as well as a rather rapid 'poo-poo-poo-poop', and birds may sit and call for long periods. Larger than White-eared Brown-dove with a noticeably heavier bill, the nape and upper back are iridescent purple, the pale stripe on the face is buff and the undertail-coverts are rich cinnamon.

YELLOW-BREASTED FRUIT-DOVE *Ptilinopus occipitalis* 29cm

Although commonly hunted, this brilliant fruit-dove remains reasonably common wherever there is primary or secondary forest from the lowlands to 1,800m. It is found throughout the country except for Palawan, occurring singly, in pairs or small flocks in the mid- to upper canopy where it feeds on a variety of fruits. The call is a single loud 'hoot' which, if constantly repeated, becomes more drawn-out with a rolling quality. Basically green on the upperparts and grey on the underparts, it has a bright yellow breast-patch bordered below with a crimson patch; the sides of face and nape are maroon and the crown is light grey.

FLAME-BREASTED FRUIT-DOVE *Ptilinopus marchei* 38cm

One of the most beautiful pigeons in the world, this large uncommon fruit-dove is endemic to Luzon, where it is found in the high oak forests of the main mountain ranges above 1,000m. Calling rarely, a deep hoot, just after dawn, the bird is hard to see and often most easily detected by its noisy wing-beats as it flies from its roost to feed. It stays within the canopy of the forest, feeding at fruiting trees. The crown and nape are crimson, the face-patch black, the upperparts black glossed green with yellow edges to the primaries and a grey band on the end of the tail. The underparts are grey with a bright orange throat and red upper breast.

BLACK-CHINNED FRUIT-DOVE *Ptilonopus leclancheri* 25–28cm

Like the Reddish Cuckoo-dove, this would be a Philippine endemic but for the population on Lanyu Island off Taiwan. It occurs singly and in pairs through most islands, in lowland primary and secondary forests. Mount Makiling is a good place to find it, but it keeps high up under the canopy and is hard to see within the green vegetation. The call is a loud 'woou', stressed on the first half. The head, neck and breast are a pale grey contrasting with emerald-green upperparts and a blackish-maroon band on the lower breast. The rest of the underparts are green except for the cinnamon undertail-coverts. It has a black spot on the chin. Females have grey replaced by green.

GREEN IMPERIAL-PIGEON *Ducula aenea* 44cm

This is a large pigeon, still reasonably common in undisturbed forest. Strictly a lowland bird, it is found on the edge of forest and more open country with scattered trees. In the early morning, small flocks gather on dead snags atop tall trees before moving off to feed in some large fruiting tree, such as a fig or mangrove, often with other pigeons and hornbills. The species has a variety of calls, usually deep growling mournful notes such as 'wrrook-rrroo' or 'wooo-wooo'. The head, neck and underparts are pinkish-grey in the adult, the nape dark chestnut and the upperparts coppery-green, blacker on the flight feathers.

METALLIC PIGEON *Columba vitiensis* 42cm

The upperparts of these large heavy-set pigeons are blackish-grey and the underparts dark grey, all washed with a metallic purple gloss. The wing-coverts, rump and tail are washed metallic green, the feathers having an iridescent fringe. The chin, throat and sides of face are light grey. This species ranges from the Sundas east to Fiji; in the Philippines it occurs in the lowlands of small islands such as the Batanes, and in the high mountain forests of the larger islands. It will travel over open country to lower altitudes in search of fruiting trees. The call is a deep two-note 'wooo-wooo', the first rising, the second lower.

PIED IMPERIAL-PIGEON *Ducula bicolor* 38cm

This large black and white pigeon is unmistakable: the entire body is creamy-white except for the black flight feathers and tail. It occurs throughout the Philippines from northern Luzon to the Sulus, roosting and breeding on small islands, often flying to the mainland to feed. It appears to be locally migratory, tens of thousands roosting on Ursula Island off southern Palawan from January to June but then disappearing for several months – probably to the south – before returning late in the year. It breeds communally with hundreds of nests placed in the canopy 10–25m up. The call is a deep 'hu-hu-hu-hu'.

PHILIPPINE CUCKOO-DOVE *Macropygia tenuirostris* 35–41cm

Medium-large with a long graduated tail, this dove is virtually endemic to the Philippines, but also occurs on Lanyu Island off southern Taiwan. It is widespread, ranging from lowland dipterocarps to mossy forest at over 2,000m, and is also seen in plantations and more open country. It flies low and fast through the understorey, singly or in pairs, and gives a loud, far-carrying 'bu bu bwow', the stress falling on the last note. It is dark brown above with rich lilac reflections on the upper back and neck. Head and underparts are bright cinnamon-rufous, paler on chin and throat. Females lack the gloss and have dark barring on breast and back.

SPOTTED DOVE *Streptopelia chinensis* 30cm

Once a rare winter visitor to Palawan only, this is now a common national resident, and probably the reason for the decline of the Island Turtle-dove *Streptopelia bitorquata*. Found in open country, cultivated lands, wetlands and elsewhere, the bird and its call, a melodious 'ter-kuk-kerr', are familiar to everyone. It feeds on the ground either singly or in small groups, and rises when flushed with a noisy clatter of wings. The nest is built low down in a small tree or bush. Brown above and pinkish below, it has a grey crown and broad black collar with white spots. The tail is tipped white to give a punctuated white band.

ZEBRA DOVE *Geopelia striata* 22cm

Common and widespread, in all types of open country (even housing estates), this is the only small long-tailed dove in the Philippines, where it was probably introduced from escaped captives; its range otherwise is from Malaysia to Australia. Its soft calls include a pleasant 'kurrrr cu cu', and a series of rolling notes, 'croo croo'. It feeds on grains, often in large numbers during rice harvests. It has a bluish-grey front of head and throat, brown upperparts with dark barring on the back, wing-coverts, rump, sides of neck and flanks. The breast is pinkish-fulvous, belly and vent white, tail brown edged white.

▲ Male. ▼ Female.

EMERALD DOVE *Chalcophaps indica* 25cm

This is a small ground-feeding dove, common throughout lowland Philippines in primary and secondary forest and open country with scattered clumps of trees. Although seen at dawn feeding in the open, it is generally retiring and usually detected either by its call, a low mournful 'tik-wooo', or as it flies low and fast swerving through the forest and displaying the two distinct grey bands on its back. The wing-coverts are bright emerald-green and flight feathers and tail are black in the male; the rest of the body is vinaceous except for a grey crown and white forehead, supercilium and carpal patch. The female is browner.

MINDANAO BLEEDING-HEART *Gallicolumba criniger* 26cm

This species appears to be much rarer than its Luzon counterpart and is found on Mindanao, Basilan, Dinagat, Leyte, Samar and Bohol. Rajah Sikatuna National Park on Bohol is one of its last strongholds. It resembles the Luzon bird but its red breast-spot is larger and darker, the underparts are cinnamon, and the top of the head, neck and sides of breast are metallic green. Like the Luzon bird, it has three grey wing-bars alternating with chestnut, and the calls are a close match. It prefers fairly open lowland forest with rocky outcrops and embankments on which it clambers about, and it uses an exposed rock for courtship.

LUZON BLEEDING-HEART *Gallicolumba luzonica* 25cm

The Puñulada, as it is known locally and which means 'stabbed with a dagger', is a secretive bird of primary and secondary forest on Luzon, Polillo and Catanduanes, generally not uncommon but scarce in some areas due to over-trapping for the pet trade. It feeds on the ground and scurries away when danger threatens, but will also approach someone imitating its call, a low mournful 'woooo'. Its English and local names come from the bright red patch on the white breast. Otherwise, the upperparts are generally grey with a purplish gloss, brown towards the rear and white underneath.

MINDORO BLEEDING-HEART *Gallicolumba platenae* 26cm

This pigeon is rare and little known, being restricted to the lowland forests of Mindoro and severely threatened by continuous encroachment of agriculture and logging. One known site is the lowland forest at Sablayan Penal Colony in western Mindoro and here it is trapped for the pet trade. It differs from the Luzon Bleeding-heart in having purplish-chestnut back and wing-coverts, green confined to the head, nape and hindneck (suffusing the mantle), chestnut wings with white and grey patches, white underparts with a small orange patch in the breast, and cinnamon-buff abdomen and undertail-coverts.

NICOBAR PIGEON *Caloenas nicobarica* 34cm

This fairly large dark pigeon might be mistaken for a megapode as it feeds at dusk on fallen fruit and seeds on the ground. If disturbed, it flies noisily up onto a horizontal branch in the low canopy, remaining perched as it does during the day. Although fairly widespread from the Andamans east to the Solomon Islands, it is generally rare throughout the Philippines, and is usually found on small, forested, predator-free islands, flying long distances between them in search of fruiting trees. It is also found on larger islands in undisturbed coastal forest. It is slate-grey on the head, neck and breast and metallic green on the upperparts; its two most characteristic features are its long neck plumes or hackles, slate-grey with purple/green iridescence, and its very short pure white tail. The bill is black and the feet are dark red. Immature birds lack the neck hackles and the tail is dark. The call is a raspy growl.

PHILIPPINE COCKATOO
Cacatua haematuropygia 31cm

This small endemic cockatoo is immediately told by its colour and loud squawking. It is all white except for pinkish-red undertail-coverts, yellow inner webs of the tail and flight feathers, and yellow-vermilion wash on head and crest. Once common throughout most of the Philippines, in lowland forest and open country where noisy flocks used to raid corn and rice fields, excessive trapping has now reduced it to small remnant populations. It is commonest on Palawan, where it can often be seen on Rasa Island, where good numbers gather to roost. It nests high up in large holes in tall dead trees, the site being used year after year.

BLUE-CROWNED RACQUET-TAIL *Prioniturus platenae* 23cm

This parrot ranges throughout the Philippines, except Palawan. It is now rather scarce, favouring lowland forest, usually feeding in the canopy and flying in small flocks, often at great height, over more open country in search of fruit. It is first detected by its noisy calls, a variety of screeches and screams given both in flight and when perched. It nests in cavities in large trees, often in small colonies. Like all other racquet-tails, the short tail is terminated with two long bare shafts ending in black spatulate racquets. The male is green with a pale blue crown and nape; the female has shorter spatules.

BLUE-NAPED PARROT *Tanygnathus lucionensis* 32cm

This large noisy parrot occurs all over the Philippines but, although once common, it has deserted many areas owing to relentless trapping for the pet trade. However, it can still be found on Palawan and at Subic Bay. It occurs in lowland forest and forest edge, usually nesting in cavities of large trees and feeding on fruit. Its flight is a distinctive rapid flapping with downcurved wing-tips. It is overall yellowish-green with a large red bill, brilliant green head and sky-blue hindcrown and nape. The key feature is the mottling on the wings created by gold-and-blue edging to the black coverts.

COLASISI *Loriculus philippensis* 15cm

One of the Asian hanging-parrots, this is the smallest and commonest psittacine in the Philippines, found everywhere except Palawan in lowland forest and even plantations and gardens. In general, it is green with scarlet forehead, rump, tail-coverts and throat-patch, golden-yellow crown and nape-patch, and red bill, but there are many distinct races which vary in the extent of red, yellow and orange on the head and nape. Found singly and in pairs, it feeds on flowering and fruiting trees, clambering about often upside down. In its fast undulating flight it gives a sharp twittering whistle.

GUAIABERO *Bolbopsittacus lunulatus* 16.5cm

This small chunky parrot is a Philippine endemic, found on Luzon, Samar, Leyte and Mindanao. It lives singly, in pairs or small flocks in lowland forest clearings, feeding on fruit such as figs and guava, keeping well hidden within the green leaves. The call is a loud 'zeet' or 'zeet zeet'. Males are generally bright green but yellower on the lower back; the face and nuchal collar are bright pale blue, the primaries are black with blue edges, the tail blue-green, the large bill black. Females have less blue on the face and the lower mandible is whitish.

PHILIPPINE HAWK-CUCKOO *Cuculus pectoralis* 29cm

Endemic to the Philippines, this cuckoo is found throughout the islands, although it is seldom seen on Palawan. It closely resembles the Besra *Accipiter virgatus*, being grey above and white below with light rufous on breast and belly, plus three to four dark bars, a black terminal band and rufous tip to the tail. It ranges widely from the lowlands to 2,000m, in primary and secondary forest and forest edge, but is difficult to see owing to its retiring nature. If its call is played back it will fly in and over the observer at great speed and hide again. The call is a frantic series of high-pitched 'wheet' notes each louder and higher than the preceding, followed by lower notes trailing off.

RUSTY-BREASTED CUCKOO *Cacomantis sepulcralis* 23cm

Ranging from southern Thailand to the Philippines, Sulawesi and the Moluccas, this species is sometimes 'lumped' with Brush Cuckoo *C. variolosus* of Australasia. It resembles the Plaintive Cuckoo *C. merulinus* but has all-rufous underparts except for a grey upper throat and sides of face. The upperparts are dark bronzy-brown and the tail is bluish-black with white tips and notches on the outer upper webs. It is solitary and shy, keeping to the high canopy of forest, mainly in the mountains. It has two distinct calls, the first a monotonous repeated series of 4–10 steady 'weet' notes, the second a three-note phrase repeated five times or so, becoming steadily higher in pitch. Birds sometimes call all night.

CHESTNUT-BREASTED MALKOHA *Phaenicophaeus curvirostris* 46cm

This graceful Sundaic cuckoo is restricted in the Philippines to Palawan. It lives in primary and secondary forests in the lowlands where it can be seen – most often in pairs, but sometimes in small family parties – clambering in the dense foliage and vines below the canopy of large trees. It is fairly large but slender with a strongly arched pale green bill, bare red face bordered by grey, greyish-brown crown and hindneck, metallic green upperparts and long, mostly chestnut, tail. Below, the throat is rufous becoming a darker chestnut on the breast and belly. The call is a deep 'tok-tok tok'.

PHILIPPINE COUCAL *Centropus viridis* 42cm

Common throughout the Philippines except Palawan and the Sulus, this weak-flying cuckoo occurs in the lowlands up to around 2,000m. While the Lesser Coucal *Centropus bengalensis* occupies open marshland, the Philippine Coucal prefers dense grass, undergrowth and vine-tangles in forest edge, tree-dotted scrub, plantations and bamboo groves. It has various calls, often given atop a bush in the early morning; many are loud, including a repeated 'cha-gook' and a descending 'coo-coo-coo-coo' in a variety of forms. The nominate race, found everywhere except Mindoro and the northern islands, is black glossed green with dark chestnut wings. The other races lack the chestnut wings.

ASIAN KOEL *Eudynamys scolopacea* 40–44cm

This largely frugivorous cuckoo ranges from India to Australia and occurs throughout the Philippines in lowland forest and second growth, commonly also present on small forested islets adjacent to larger islands. The call is a loud persistent 'ko-el ko-el ….' and a strident bubbly 4–6-note 'wrreep wrreep …'. It lays its eggs in the nests of crows and mynas. Although there are many subspecies, the basic pattern is consistent: large with a long tail, the male is black and the female is brown and buffy, streaked and spotted with black, brown and rufous.

▲ *Negros race.*

PHILIPPINE SCOPS-OWL *Otus megalotis* 23–28cm

This owl is strictly nocturnal and difficult to locate owing to its very infrequent calls, which consist of two or three descending growling notes. Fairly common throughout the Philippines except Palawan, there are three distinct races that differ in size, coloration and call. In Luzon, it is fairly large for a scops with prominent ear-tufts. Generally brown or mottled brown in plumage, it has an obvious pale V-shaped stripe over each eye, from bill to tufts, and a distinct pale buff wing-bar and collar. It is found in primary and secondary forest up to around 1,500m.

PHILIPPINE EAGLE-OWL *Bubo philippensis* 51cm

This, the largest owl in the Philippines, is not uncommon in primary lowland and mid-montane forests, sometimes near rivers, but it is difficult to see and remains little known owing to its nocturnal habits and infrequent calling. Of several calls the commonest is a deep bubbly 'bo-bo-bo-bo'. Very large with prominent ear-tufts, it has powerful legs for catching vertebrate prey. There are two races; one on Luzon and Catanduanes with rufous-based dark-streaked upperparts, and one on Mindanao, Samar, Leyte and Bohol with brown-based dark-streaked upperparts. The underparts are buff streaked with brown.

AUSTRALASIAN GRASS OWL
Tyto longimembris **40cm**

Widespread from Nepal to Australia, this species is sometimes treated as conspecific with the African Grass Owl *T. capensis*. It is fairly common throughout the archipelago up to around 1,750m, usually seen quartering over the fields at dusk or dawn. During the day it roosts in dense grass. The upperparts are dark brown with rufous-buff markings and some white spotting, and the underparts are white with a buffy wash. The flight feathers and tail are barred. In flight the head appears huge, dominated by its heart-shaped white facial disc. The call is a loud screech.

GIANT SCOPS OWL *Mimizuku gurneyi* **31–38cm**

Endemic to the Philippines, this remarkable owl is confined to Mindanao, Dinagat and Siargao. It is restricted to forest and occurs from the lowlands up to at least 1,500m. Strictly nocturnal, it is rather shy and often difficult to see. The usual call is an explosive, shrieking 'kiarrr', given singly or a few at a time. Also gives a single, loud, shrill, whistled 'teeoou', recalling a frogmouth, sometimes in an extended series for hours at a time. The upperparts are rufous or brown, well streaked and with prominent ear-tufts. The underparts are rufous-buff with bold black streaks. The bill is pale greyish, the eye is brown and the feet are dull pale fleshy. Juveniles are rich ginger all over, lacking streaks.

PHILIPPINE FROGMOUTH *Batrachostomus septimus* 23cm

This endemic night bird occurs locally on most larger islands except Palawan and Cebu, in forests and second growth up to 2,000m or more. By day it perches upright on branches with eyes closed. It has a broad stout yellowish bill with a very wide gape, but rather than hawk for insects it picks them from the ground and branches. The nest, a pad of feathers on a horizontal branch, holds a single white egg which the adults incubate sitting very erect. The call is a harsh descending growl plus a loud screaming chatter, 'wak-wak-wak-wak…'. The species occurs in two phases, brown and rufous. In the former the underparts are light brown with two white cross-bands; the upperparts are mottled grey-brown with a white collar and white spots on the wing-coverts. In the latter the brown is replaced by rufous and the bird is less mottled.

SAVANNA NIGHTJAR *Caprimulgus affinis* 20cm

Widespread from India to South China and Lesser Sundas, this night bird is patchily distributed in the Philippines but is often common. It is found on dry riverbeds, open stony country, shingle and barren grasslands, even in the suburbs of Manila. It forages low over forest, cultivation and settlements, snapping up insects drawn to artificial lights. The call is a high-pitched, far-carryng 'chiveet', constantly repeated for long periods. The nest is a slight scrape on bare ground in which one or two pinkish or buffish eggs are laid. This is a smallish, generally rather uniform nightjar with a relatively short tail. It is sand-brown above and below, with a vermiculated pattern of black spots and bars. The male has small white patches on the side of the throat and tips of the outer tail feathers. The female lacks the white on the tail, which has four wide buffy bars.

WHISKERED TREESWIFT *Hemiprocne comata* 16cm

This bird usually occurs on exposed branches atop tall trees at the forest edge from where it flies out, in broad arcs, to hawk for insects such as wasps and beetles. It returns to the perch folding its wings in neat jerks. It has a loud high-pitched clear twitter. The nest is a small shallow cup attached to a branch, and holds a single white egg. The head, throat, tail and wings are a dark metallic blue, the belly is white, the rest a bronzy-brown; two long white stripes above and below the eye are diagnostic. The male has a chestnut ear-patch. The tail is deeply forked and the wings fold across the body and below the tail.

GLOSSY SWIFTLET *Collocalia esculenta* 9.5cm

Swiftlets are small swifts that possess a slower, less regular flight than their larger relatives. The Glossy or White-bellied Swiftlet is common throughout the Philippines at all altitudes, over forest, open country and seashores, occurring in five races; one, the grey-rumped marginata, is sometimes regarded as a separate species. The upperparts are a glossy blue-black whilst the chin and throat are grey, becoming whiter towards the belly. It breeds in small colonies often under eaves or in buildings, cementing its half-cup nest of moss and lichens to a sheltering wall.

COMMON KINGFISHER *Alcedo atthis* 15cm

Found from Europe to Japan, this brilliant bird is a common migrant to the Philippines, arriving in September and occurring throughout the archipelago. It can be found on the coast, along streams and rivers, in fishponds and marshlands, diving for fish from a low waterside perch or hovering over the water before plunging after its prey. It flies low over the water with a high pitched whistled 'cheee'. The head and nape are banded dark blue-green and blue, the back and rump are light aquamarine and the wings and tail are dark blue-green. It has a rufous band behind the eye and a white patch on the side of the neck. The broad malar stripe is barred blue-green and black. It is rufous below with a white throat. The similar Indigo-banded Kingfisher *Alcedo cyanopectus* is smaller and darker with one or two breast bands, depending on sex.

VARIABLE DWARF KINGFISHER *Ceyx lepidus* 14cm

This small solitary kingfisher perches inconspicuously in the undergrowth of primary and secondary forests, sallying out to catch insects. It ranges from the Moluccas to New Guinea, but is local in the southern Philippines, commoner on smaller islands such as the Sulus and Camiguin. The call is a shrill high-pitched 'peteet'. The upperparts vary from metallic blue to blackish and sometimes silvery-white, with blackish wings and tail. The lores, cheeks and most of the underparts are orange-rufous; the throat and mid-belly are white, with a white spot on the neck. The bill and legs are orange-red.

SILVERY KINGFISHER *Alcedo argentatus* 15cm

This beautiful little kingfisher is endemic to the southern islands of the country from Basilan to Bohol. It is not uncommon by streams in or near lowland forest, flying very fast along them with a high-pitched 'wheeet'. It particularly favours pools made by dammed streams where it will perch on an exposed trunk or branch. Its black and white plumage is unmistakable: the face, upperparts and wings are black, the mid-back, rump and tail-coverts silvery-white. The throat and belly are white and breast-band, flanks and undertail-coverts black with a bluish cast. It has bright orange-red legs and a black bill.

PHILIPPINE DWARF KINGFISHER *Ceyx melanurus* 12–13cm

This bird has the same habits and habitat as the Variable Forest Kingfisher, but also occurs on Luzon, Leyte and Samar. Its presence is often first revealed by its high-pitched whistle as it takes flight. It nests in embankments in the forest, often in a dry gully. The Luzon race is the smallest, rufous above spotted lilac. The wings are black, lightly spotted blue, while the scapulars are part-black, part-rufous, forming two bands. A white line along the neck has a blue spot above it. The upper throat and mid-belly are white, the breast and flanks lilac-rufous.

COLLARED KINGFISHER *Halcyon chloris* 24cm

This kingfisher, which ranges from Africa to China and south to Australia, is the commonest in the Philippines, only absent in deep forest and high mountains. Fond of coastal areas, mangroves and exposed reefs, it sits on stumps, rocks or other suitable perches looking for crabs, shrimps and fish at low tide, and also on telegraph wires near paddies looking for frogs. It is generally turquoise above, slightly greener on the upper back and crown, and white below. It has a broad white collar bordered black above. It is noisy, announcing its presence with a loud harsh 'kak-kak-kak-kak'. It nests in holes in trees, embankments and even termite nests.

STORK-BILLED KINGFISHER *Pelargopsis capensis* 33cm

This Asian kingfisher with its massive scarlet bill is the largest in the Philippines but patchily distributed. Birds on Palawan and Mindoro have the whole head, neck and underparts cinnamon-buff, the back and rump aquamarine and the wings and tail greenish-blue. Elsewhere the rufous parts are paler, the wings and tail greener; in the Sulus the head and underparts are buffy-white. The species is found in more secluded coastal and riverine areas, feeding on crabs and fish. The call is a loud, harsh 'kak-kak-kak'.

RUDDY KINGFISHER *Halcyon coromanda* 28cm

Four races of this bird exist in the Philippines, two resident and two migrant. The migrants come from Japan and the Ryukyu Islands, dispersing thinly in winter. The residents are found on Palawan and the Sulus, where they are not uncommon along forest streams and in mangroves and beach forest, but are hard to see as they sit quietly in the understorey from where they drop to the ground to catch frogs and lizards. The upperparts are rufous with a lilac wash, the lower back and rump bright bluish-violet. The underparts are rufous-ochre, darker on the breast; the upper throat is white, the bill red. The call is a delightful series of descending whistles.

WHITE-THROATED KINGFISHER *Halcyon smyrnensis* **26cm**

This fairly large kingfisher is found commonly throughout southern Asia and the Philippines except Palawan. It occurs almost anywhere except in deep forest, but is most usually observed sitting singly on bare branches and telegraph wires near water in cultivated areas, plantations and forest edge, diving down for a wide variety of food, commonly fish and lizards. It nests in tunnels burrowed in riverbanks and ants' nests. The head, mantle and underparts are chestnut, the throat white, the back, uppertail and secondaries turquoise, wing-coverts, undertail and primaries black, with a white patch at the base of the primaries obvious in flight.

SPOTTED WOOD-KINGFISHER *Actenoides lindsayi* 25cm

This large kingfisher, endemic to Luzon, Negros and Panay, occupies lowland and mid-montane forest. It is very vocal at dawn, giving a sharp whistle, 'tuooo', a series of descending notes led by a trill, and an agitated chatter. Otherwise it is secretive, sitting silently and dropping to the ground for insects and snails. The male is dark green above with buff wing-spots, white below with green-margined breast feathers. Three bands circle the crown: blue above the eye, black below the eye, then rufous forming a collar. The malar stripe is blue, the throat cinnamon. In females the blue is replaced by green.

BLUE-CAPPED KINGFISHER *Actenoides hombroni* 28cm

This Mindanao endemic lives deep in primary and secondary forests mostly in the mountains, but seems rarer where it competes with the Rufous-lored Kingfisher. It perches in the understorey, feeding on snails and frogs. Its calls include a staccato chatter, whistled 'tiyu' and a stuttering trill with drawn-out single notes. The crown, nape and malar stripe are deep blue, neck and collar rufous, mantle and wing-coverts blue-green spotted ochre, lower back and rump light blue, and tail dark blue. The throat and abdomen are white, breast and flanks tawny. Females have crown, malar stripe and tail green.

RUFOUS-LORED KINGFISHER *Todiramphus winchelli* 24cm

Endemic to the Philippines where it ranges from the Visayas to the Sulus, this is an uncommon bird of primary lowland forest, often perching just below the canopy. It feeds on small animals and insects on the forest floor. Its call is a series of loud ringing notes that get higher and louder, sometimes ending with a set of descending notes. It is blackish-blue above with a light blue crown-band. The lores and nuchal collar are rufous, the lower back and rump silvery-blue. The male's underparts are all white; the female has throat, breast and flanks washed rufous.

BLUE-THROATED BEE-EATER *Merops viridis* 28cm

Bee-eaters, found throughout the warmer parts of the Old World, have long pointed bills for catching bees and other flying insects, swooping down on them from exposed perches like telephone wires and bare branches. This species is found in colonies of 6–20 or more, usually in lowland forest but also in more open part-wooded country, nesting in holes in sandy embankments and even level ground. It has elongated central tail feathers, the crown to upper back is chestnut (deep green in the juvenile), the cheeks, lower back and tail are blue, the wings green tipped black, and the underparts pale lime-green.

BLUE-TAILED BEE-EATER *Merops philippinus* 29cm

This common but local bee-eater, which ranges from South Asia to New Guinea, is found throughout the Philippines in more open country than the Blue-throated, preferring wetlands, ricefields and cultivations with scattered bushes. It perches on small trees and telegraph wires, and flies with lazy circular glides, catching insects in the air. It is usually in small groups. The yellow chin and chestnut throat are diagnostic. The lores and mask are black outlined with blue, the head and mantle green, the rump and tail blue, the breast and belly green becoming light blue on the undertail-coverts. The call is a liquid 'chrrrip'.

DOLLARBIRD
Eurystomus orientalis 25cm

The Dollarbird or Broad-billed Roller is a chunky medium-sized bird that sits conspicuously for long periods on exposed branches on tree-tops, at the edge of lowland forest or in more open country with many old trees. It is more active in the early evening when it flies around in a jerky, buoyant manner, hawking for insects; the silver-blue 'dollars' at the base of the primaries can then be seen. The nest is a hole in a tree. The head is black, the rest mostly bluish-green except for blue-edged black flight feathers, black tail and blue-streaked black throat. The broad bill is red, the short legs orange.

LUZON HORNBILL *Penelopides manillae* 54cm

This small endemic hornbill is found on Luzon, Catanduanes and Marinduque. It can be fairly common in forested areas up to 1,500m, and is often found in pairs or noisy parties. The call is an abrupt, squeaky, nasal 'eenk', like a cheap toy trumpet. The nest is a cavity sealed by the male in the trunk of a large tree; the female receiving food through a narrow opening. The male has dirty yellowish-white head and underparts, with the throat, ear-coverts and upperparts blackish with a greenish gloss on the back. The tail is whitish with a broad black terminal band. The female is overall blackish. Similar species occur on other islands throughout the Philippines (with the exception of Palawan) but some are very rare.

WRITHED HORNBILL *Aceros leucocephalus* M 77cm, F 69cm

On Mindanao and its satellite islands Camiguin Sur and Dinagat, to which it is restricted, this medium-sized hornbill is not uncommon in undisturbed lowland forest, sometimes forming large flocks especially in fruiting fig trees or when going to roost. Rampant deforestation and hunting have seriously affected the population. Its breeding habits are similar to the Tarictic and other hornbills. The call is a single, relatively soft, bleating 'onk'. The male is a glossy greenish-black, the head and neck yellowish-buff and the crest chestnut. The tail is white with a black tip. In the female, the head and neck are black.

WALDEN'S HORNBILL *Aceros waldeni* M 79cm, F 69cm

Until recently, this extremely endangered hornbill, endemic to Panay and Negros, was lumped with Mindanao's Writhed Hornbill, but among its distinguishing features is its unique call, a loud nasal honking 'au au au auk'. It lives in small groups but gathers in larger numbers at fruiting figs, and is found in the canopy and edge of the tiny and rapidly diminishing primary lowland forest left within its range. The head to upper breast are dark rufous (females have a black head and neck), breast and belly black, back and wings black with green gloss, and tail buffy with black terminal and basal bands.

RUFOUS HORNBILL *Buceros hydrocorax* M 94, F 89cm

This is the largest and most spectacular of the Philippine hornbills, with three distinct races. It travels in family groups in forest up to 1,500m, gathering in fig and other fruiting trees to feed. Its loud 'kaa-ow' can be heard from afar, and is given in flight, on alighting and when feeding. It nests in large tree cavities in which the female is sealed and fed by the male. In Luzon, the head and neck are chestnut and the huge bill red with a flattened casque. The breast is black, the belly chestnut, mantle and wings brown, and tail white stained yellowish-buff.

COPPERSMITH BARBET *Megalaima haemacephala* 15cm

This common Asian bird ranges throughout the Philippines except Palawan and the Sulus. Revealed by its monotonous call, a long series of resonant 'pok' notes, it inhabits forest and more open country and nests in holes in trees. Although usually seen singly or in pairs, often atop high dead limbs, larger numbers gather at fruiting trees. The forecrown is crimson, with a black midcrown and blue-green nape. The ear-coverts, throat and breast are yellow with a crimson band across the upper breast; the belly is pale green with dark streaks. West Visayan birds have crimson throats and ear-coverts.

SOOTY WOODPECKER *Mulleripicus funebris* 35cm

This spectacular woodpecker is endemic to the Philippines. The northern race, *funebris*, found on Luzon and related islands, is sooty-black, the male having a red forehead, face and malar stripe. The southern race, *fuliginosus*, is found on Mindanao, Samar and Layte, and is paler and greyer with red in the male confined to the malar stripe. In both races the eye is pale yellow and the bill ivory-white, and the female lacks red on the head. Usually travels in pairs and perches conspicuously on dead trees and snags. Found in primary forest with many large trees, it attracts attention with its loud drumming and its call, which recalls a shrill referee's whistle.

WHITE-BELLIED WOODPECKER
Dryocopus javensis 40cm

This large Asian woodpecker is found singly or in pairs throughout the Philippines in lowland primary forest. It has a powerful bill for excavating nest holes in trees and for uncovering bark-living insects. Like all woodpeckers, it clings to the trunks and branches of trees, using its stiff tail as a prop. It is black with a buffy-white belly and various amounts of white on the throat, head and breast. In males, the crown, nuchal crest and malar patch are bright crimson; in females, only the crest is crimson. The call is a sharp yelping 'kiyow'.

PHILIPPINE PYGMY WOODPECKER *Dendrocopus maculatus* 14cm

Endemic to the Philippines and found at all altitudes in forest, second growth and clearings throughout the islands, this small woodpecker feeds on insects and larvae while creeping along the thinner trunks and outer branches of trees, and often favours dead limbs. It is found alone or in pairs but generally occurs in mixed feeding flocks with other small birds. There are many geographical variations, but birds are generally black or brown, barred and spotted with white. The underparts are yellowish or buffy-white, lightly streaked with black or brown. The call is a far-carrying, stuttering trill.

VISAYAN WATTLED BROADBILL *Eurylaimus samarensis* 16.5cm

Broadbills are a small Afro-Asian family with large heads, heavy broad bills, short legs and usually colourful plumage. This species is found on Leyte, Samar and Bohol, while the closely related Mindanao Wattled Broadbill *E. steeri* is found on Mindanao, Basilan, Malamaui, Dinagat and Siargao; both in lowland primary and secondary forest. Similar in appearance, with a striking, broad, pale blue wattle around the eye. The Visayan form has a white and lilac bar in the wing and a greyish collar, the Mindanao a white and yellow wing-bar and clear white collar. Males are pinkish-lilac below, females white. Found singly or in small parties, often in mixed-species flocks. Hard to spot, although may attract attention by the whirr of its wings.

WHISKERED PITTA *Pitta kochi* 21.5cm

Pittas are plump, short-tailed, colourful but secretive birds which hop on the forest floor turning over leaves for invertebrates. Once thought rare, this magnificent species, endemic to Luzon, has proved locally common in the high mountain oak forests of the northern Sierra Madre. Its call is a series of deep hollow notes, the first rising, the others descending, given with bobs of the head from a perch in the understorey. The key features are its pinkish 'whisker' and brilliant scarlet belly; otherwise it has a brown face, rufous hind-crown, olive-brown upperparts and greyish-blue tail, wing-coverts and breast-band, with white-spotted black flight feathers.

HOODED PITTA *Pitta sordida* 16.5cm

This bird, also known as the Green-breasted or Black-headed Pitta, is a fairly common resident in the Philippines, found in dry open forests and scrubland with scattered patches of trees or in vegetation along creeks, even venturing into bamboo groves near habitation. Although naturally retiring, it responds readily to a playback of its call, a loud double note, 'waup-waup', which it gives regularly in the rainy season, usually from a branch. It builds a large round nest on the ground covered with vegetation. It is basically dark green with a black head and throat, cobalt-blue wing-coverts and rump, black patch on belly and crimson vent. When flushed, its white wing-patches against dark wings are very obvious. There are two generally recognized subspecies in the Philippines, the widest-ranging one being on all major islands except Tablas, Catanduanes, Polillo and Palawan; on the latter island and its satellites the birds have a smaller black belly-patch and are lighter green with a deeper blue rump.

STEERE'S PITTA *Pitta steerii* 20cm

This bird, much pursued by birdwatchers owing to its beauty and mystique, is endemic to Mindanao, Bohol, Leyte and Samar, where it is known from only a handful of sites; Rajah Sikatuna National Park in Bohol is an excellent place to see it. It inhabits forests below 1,000m and appears to favour areas with limestone outcrops. The song consists of 5–6 forceful 'werp' notes, often given from a tree. The head, neck and tail are black, the back and inner secondaries green, wing-coverts and rump metallic-blue. Below, the throat is white, breast and flanks are sky-blue, belly black and vent scarlet.

BARN SWALLOW *Hirundo rustica* 18cm

One of the best-loved birds in the world, breeding in the northern hemisphere in North America and Eurasia and wintering in tropical areas, the Barn Swallow is common in the Philippines from September to May in all types of habitats, but most frequently near water, with huge roosting flocks forming in reedbeds. It builds an open mud nest fixed to buildings, but spends much time on the wing. Its long forked tail with elongated outer feathers and red throat and front are diagnostic; the tail has white spots above and below. The upperparts and breast-band are steely-blue, the belly and underwing range from white to rufous.

HOUSE SWALLOW *Hirundo tahitica* 13cm

Sometimes known as Pacific Swallow, this is a common resident throughout the Philippines, living in cultivated areas and near human habitations often near water. It is commonly seen in large numbers hawking insects over ricefields, and its habits are similar to the Barn Swallow, from which it is told by its smaller size, shorter tail, brownish-grey underparts and lack of dark breast-band. The breast and throat are rufous, the upperwings and tail black, the latter with white subterminal spots, the underwing uniform dusky brown. The call is a soft twitter. The nests are half-saucers of mud fixed to buildings and bridges.

PIED TRILLER *Lalage nigra* 16cm

Common throughout the Philippines in forest edge, scrubland, plantations, mangroves and gardens, the Pied Triller occurs in pairs or loose flocks, feeding on insects in small trees. It flies with a slow undulating flight, sometimes coming to ground, and calls with a rapid ascending 'kek-kek-kek'. The male has a glossy black crown and upper back, grey lower back and rump, black tail with white tips, and white underparts, wing-bar and wing-covert fringes. It is told from the larger Black-and-white Triller *Lalage melanoleuca* by its white eyebrow and black eye-line. Females are duller, greyer and lightly barred below.

COMMON IORA *Aegithina tiphia* 13.5cm

This pretty Asian bird is restricted in the Philippines to Palawan where it is found in beach forest, mangroves, tree-dotted country, cultivations and gardens. It is not easy to see as it feeds high in the trees, but it has a distinctive harsh accelerating stutter with a final sharp note, 'chi-chi-chi-chi-chi-pow'. Males have black wings, a white shoulder-patch and two wing-bars. The forehead, throat and breast are bright yellow, the upperparts olive-green, yellower on the crown. The tail is black, the flank feathers long, fluffy and white. The female is duller with an olive-green tail and paler wing markings.

BLACK-HEADED BULBUL *Pycnonotus atriceps* 17cm

This small colourful bulbul is distributed from India to Bali but is restricted in the Philippines to Palawan where it is fairly common in open country with scattered trees, mangroves and forest edge. Usually in small groups, it likes to perch out in the open on exposed limbs. Its call is a pleasant chirp as it flies actively between the trees, and its song is a series of mournful 'chip chip' notes. It is easily identified by its glossy black head and throat, olive-yellow plumage with black flight feathers, and bright yellow tail with black sub-terminal band. The bill and legs are black, and the eyes grey-blue. It feeds on fruit and berries. The nest is compact and bulky and placed between upright forks of branches.

YELLOW-VENTED BULBUL *Pycnonotus goiavier* 19cm

This cheeky bird is one of the most abundant in the Philippines, occurring anywhere except in primary forest, from urban gardens to open country up to 1,800m, and is one of the few birds to be found in Metro Manila. It probably owes its success to its omnivory, since it eats not only fruits and berries but also worms and insects on the ground. In the evening it often gathers in large numbers at communal roosts. Its call, as its local name implies, is 'cul cul'; it also has a variety of chuckles. Generally brown above and white below, it can be identified by its yellow vent and white supercilium over black lores.

SULPHUR-BELLIED BULBUL
Ixos palawanensis 17cm

This small, fairly common bulbul is found only on Palawan where it is usually seen in small noisy flocks feeding on vines and in the dense vegetation in and on the edge of lowland forest, often quite close to the ground. It gives various chip-like calls. It somewhat resembles the larger Grey-cheeked Bulbul *Criniger bres* but lacks the latter's crest and white throat and has a thinner bill. The plumage of the sexes is similar, being olive-brown above, and yellow washed with olive on the breast and flanks. The eye is pale yellow.

PHILIPPINE BULBUL *Ixos philippinus* 22cm

This Philippine endemic, found everywhere except Palawan, is an abundant forest dweller – even in small pockets in open country – from sea level to the mountains. There are five races, some quite distinctive, which vary in plumage and song. They have long decurved bills, long wings, pointed grey crown feathers and olive-brown upperparts, the throat and breast varying between races – e.g. rufous on Luzon, greyish-brown on Mindoro – but all having white shaft-streaks and a whitish belly. The song is a short musical phrase, on Luzon 7–8 notes, descending then rising. The call is a raspy chatter.

YELLOWISH BULBUL *Ixos everetti* 24cm

Endemic to the southern Philippines from Samar to Tawi-Tawi, this bird is a noisy and conspicuous inhabitant of the canopy and edge of lowland forest, often in loose flocks at fruiting trees. In Mindanao, the call is a typical raspy 'yeeeeuk' and the song a rapid melodious series of whistles, the last two distinctly higher; the songs of the other races are quite different. The upperparts are yellow-olive with a shaggy streaked crown, the throat and chest are tawny-yellow, the belly yellow and flanks olive. The highly distinct race on Camiguin Island is larger, with a slight crest, goaty beard and massive blue-grey bill.

HAIR-CRESTED DRONGO *Dicrurus hottentottus* 25cm

This Asian-Pacific species resembles the (Luzon) Balicassiao in size and colour, the upperparts being glossed blue-green, the back bluer, with a velvet-black frontal crest. The throat and upper breast are spangled blue-green. In the south the tail is short and square whilst elsewhere, including Palawan, it is long and forked, reaching the maximum length of 185mm on Tablas. Like other drongos it is bold and noisy, with a wide range of calls including mimicry of other birds. It builds a hanging nest made of plant fibres. It is found in lowland forest and forest edge up to around 1,300m.

▶ *Black and white form.*
▼ *Black form.*

BALICASSIAO *Dicrurus balicassius* 27cm

This drongo is endemic to the northern Philippines, in two distinct forms: all black in Luzon and Mindoro, and black with a white belly in the Western Visayas. Belly apart, the black has a distinct green gloss. Luzon birds have a short velvet-black frontal crest. The black bill is strong, slightly curved and hook-tipped. It is bold, noisy and strictly arboreal, in all storeys of dense forest, in pairs or small loose groups. It is very conspicuous in dense lowland forests, calling loudly with varied whistles, rattles and cat-like 'meoows', and can be very responsive to spishing.

BLACK-NAPED ORIOLE *Oriolus chinensis* 27cm

This boldly coloured bird is common throughout the Philippines in open country, orchards, gardens and partly cleared second growth, commonly first glimpsed in its characteristic undulating flight. It often roosts in large groups. The call is a loud, fluid, variable whistle such as 'che-lee-che-leoo'. The nest is a pendant of plant fibres hung at the end of a branch. The male is mainly golden-yellow with black wings, tail, crown, nape and ear-coverts, with yellow forecrown and tail-tips. The female is similar but olive-yellow on the mantle. Young birds are yellow-green above and dark-streaked creamy below.

ASIAN FAIRY-BLUEBIRD *Irena puella* 24cm

In the Philippines this widespread Asian beauty is restricted to Palawan. Sometimes considered an oriole and sometimes a leafbird, it is a noisy, stocky bird of lowland forest, subsisting on fruits and insects and usually occurring in small groups outside the breeding season, especially at fruiting trees. The nest is a layer of twigs lined with moss and placed in the fork of a tree. Males have the crown, nape, back, wing-coverts, rump and vent a shining blue, the rest black. Females are greenish-blue with black primaries. The call is a series of sharp ringing 'whit' and 'whit whit-ah' calls.

PHILIPPINE FAIRY BLUEBIRD *Irena cyanogastra* 25cm

This species of lowland primary forest is endemic to the Philippines in four races, one on Luzon, one in the Eastern Visayas, one on Dinagat and Mindanao and one on Basilan. It is a mixture of blackish-blue and velvety black with the crown, nape, part of the wing-coverts, rump and tail-coverts bright cobalt-blue. It can be mistaken for a drongo, whose glossy plumage can appear blue but which is slimmer and has a slightly forked and curved tail. It occurs singly or in pairs, and sometimes in larger groups in fruiting trees. The call is a short series of loud snappy whistles, with a sharp whip-like note, e.g. 'hu-wee-u WHIP hu-wee-u'.

LARGE-BILLED CROW *Corvus macrorhynchos* 47cm

Crows are a family of strong-billed birds, keenly intelligent and often associated with man. They have harsh calls, make large untidy stick nests and feed on fruit and animal matter, often scavenging. The Large-billed Crow is glossy black all over with a particularly large black bill. It is seen singly or in pairs and found in open cultivated country, coconut and forestry plantations, often near villages. It ranges from Iran to China and south as far as Sulawesi, and is a common bird in the Philippines. The call is a harsh 'wark-wark'.

ELEGANT TIT *Pardaliparus elegans* 11cm

This is one of three endemic Philippine tit species, found everywhere except Palawan and occurring in many well-marked races. It is common in primary and secondary forest at all altitudes, in mixed flocks and family parties moving up and down branches, often upside down looking for insects. Adults are black on crown, mantle, throat and breast, yellow from cheek to shoulders, and bright yellow on belly and undertail-coverts. The wings are black with white blotches. Young birds are olive-green above and have brown wings and head with yellow patches. It has a variety of pleasant whistled phrases and nests in hollow trees.

SULPHUR-BILLED NUTHATCH *Sitta oenochlamys* 12cm

This bird, often treated as a yellow-billed form of the widespread Velvet-fronted Nuthatch *Sitta frontalis*, is common in lowland and montane forest throughout the country except Mindoro, Palawan and the southern islands. It is sometimes found in mixed flocks, climbing up and down trunks and branches, using its long straight bill to search for insects and larvae under the bark. It nests in hollow trees, and the call is a strong trill-like 'pip-pip-pip-pip'. It is a violet-washed soft blue above, rosy-buff below. The forecrown is black, and the male has velvet-black eyebrows.

STRIPE-BREASTED RHABDORNIS *Rhabdornis inornatus* 15cm

Once treated as treecreepers, the three species of rhabdornis are now put in their own family, endemic to the Philippines. They occur alone, in small parties or mixed feeding flocks in the canopy of lowland forest, hopping along branches in search of insects and larvae, also sometimes perching on high bare limbs. The call is a high, thin 'tzit', often repeated in series just before taking flight. Stripe-breasted Rhabdornis is found on Samar, Leyte, Biliran, Panay, Negros and Mindanao. The upperparts are brown with a narrow grey-white supercilium, while the underparts are white with soft warm brown streaks.

FLAME-TEMPLED BABBLER *Stachyris speciosa* 13cm

This beautiful bird, restricted to forest below 1,000m on Negros and Panay, is now sadly endangered owing to loss of habitat. One good place for it is Casa Roro near Dumaguete. It lives in the understorey and at the forest edge and joins mixed feeding flocks. The call is a two-second phrase of cheerful, melodious, bouncy whistles. It has a brilliant powdery-yellow bill, face and eye-ring, the crown and throat are black, the temples long and orange. It has a black and olive-yellow collar, otherwise being olive-grey streaked grey above, yellow below with black spots.

STREAKED GROUND-BABBLER *Ptilocichla mindanensis* **16cm**

Secretive and difficult to see, this species lives in low growth, feeding on insects around fallen trees in the forests of Bohol, Samar, Leyte and Mindanao to around 1,300m. Unlike other babblers, it is not gregarious but nearly always found in pairs. It has a loud strident whistle repeated two to three times at different pitches, usually followed by several discordant imitations of other birds. Very responsive to playback of its call, it runs around like a little mouse. It is a rich dark brown with pale shaft-streaks, white eyebrow and throat, and the underparts are dark brown with white stripes.

CHESTNUT-FACED BABBLER *Stachyris whiteheadi* 14cm

A rapid series of sharp 'chip-chip' notes, alternating with a repeated 'pi-chu', introduces this medium-small Luzon endemic, mainly montane but sometimes in associated foothills. Outside the breeding season it lives in groups of up to 30, often joining other birds in mixed flocks, actively moving through all levels of the forest searching for insects at a feverish rate. When flying, the wing-beat is audible. The face is chestnut and the broken eye-ring white; the crown and nape are grey, the remaining upperparts olive-green. The underparts are olive-yellow, brighter on the belly.

MELODIOUS BABBLER *Malacopteron palawanense* 20cm

This medium-large babbler earns its name from its beautiful dawn song, a series of five loud whistles that alternate high and low in pitch, often preceded by loud 'whit whit' notes from several birds at once. It is endemic to the lowland forests of Palawan where it skulks amid the dense tangles and vines; outside the breeding season it nearly always travels in small flocks. The upperparts are olive-brown except the tail and forecrown which are rufous. The sides of the head are grey and the underparts white with a brown wash on breast and flanks. The bill is long and slightly hooked.

BROWN TIT-BABBLER *Macronous striaticeps* 14.5cm

This fairly small, noisy but reclusive babbler, endemic to the islands from Samar south to Tawi-Tawi, travels through the dense understorey and edge of forest in small parties, keeping close to the ground, actively feeding. It is often seen crossing trails, flying low and weakly, and dropping immediately into cover. Several birds will call together, giving a chattering raspy 'we-chu we-chu'. It is rufous-brown above, buffy below with an off-white throat, black head and face with white shaft-streaks. Some back feathers have white shafts extending beyond the base of the tail, giving it a fluffy appearance.

WHITE-BROWED SHORTWING *Brachypteryx montana* 13–14cm

This fine Asian songster is found in forests, often near streams, on the mountainous islands of the Philippines above 1,000m. Differences in song between upper and lower populations may mean that two species are involved. One song is a series of loud warbling whistles that rise up and down; others lack uniformity. The species is very secretive and mouse-like, keeping close to the ground. It has long legs and short wings and tail; the male is a dark slaty-blue with a blackish head and concealed white eyebrow. Females show varying amounts of brown; some races have all-brown heads.

SIBERIAN RUBYTHROAT *Luscinia calliope* 15cm

Breeding in north-eastern Asia and migrating south in winter, this terrestrial robin is a common winter visitor to the Philippines, particularly Luzon. It inhabits marshlands and open country with long grass, skulking in thick tangled vegetation where it feeds on insects, but sometimes coming out onto tracks in the early morning. The call is a plaintive whistle 'chee-wee'. Both adults have prominent white eyebrows and moustachial stripes; the male has a startling red throat whilst the female's is whitish. The upperparts are brown, the underparts whitish with buffy flanks and grey on the breast.

ORIENTAL MAGPIE-ROBIN *Copsychus saularis* 20cm

The cocky magpie-robin is a familiar, fairly common bird throughout the islands although absent from Palawan. It is usually found in open cultivated country in bamboo groves, coconut plantations, small wooded areas often near farms and villages. It moves about the bushes and trees boldly, cocking and fanning its tail. Very territorial, it is also quite vocal and has a loud and varied melodious song; the call is a high-pitched whistle. The male is unmistakable with its bluish-black body, white belly and white patch on the wings. In the female the black is replaced with dark grey.

WHITE-BROWED SHAMA *Copsychus luzoniensis* 18cm

The beautiful song of this common but skulking thrush, a long series of rising and falling whistles and gurgles, gives its presence away. The head, back, throat and upper breast are bluish-black with a long white supercilium that enlarges when the bird is excited. It has a chestnut lower back and rump, black tail with white tips, black wings with a white bar, white belly and brown flanks. The rarely seen female has an olive-brown crown, grey-brown back and grey throat. The species inhabits dense undergrowth in lowland forests on Luzon and Marinduque with another race, lacking the chestnut rump and white wing-bars, on Panay and Negros.

CEBU BLACK SHAMA *Copsychus cebuensis* 20cm

Endemic to Cebu, this thrush is endangered owing to the island's near-total lack of forest, although it also uses bamboo groves where it is locally not uncommon, the bamboo-covered creeks and hillsides at Casili, Consolacion, near Cebu City, being a favoured area. Its habits are like the White-browed Shama's, and it also has a beautiful song, a series of more plaintive melodious whistles rising up and down. It builds its cup-shaped nest in broken-off bamboo stumps. Medium-sized with a long graduated tail, the male is glossy blue-black, the female grey-black. The eyes and legs are black.

PIED BUSHCHAT
Saxicola caprata **12.5cm**

This small active little bird, widespread from Iran east to China and south to New Guinea, is a familiar sight throughout the Philippines in open country, grasslands, paddies and marshes, sitting on prominent perches such as fence posts, telegraph wires, and tops of bushes and rocks. The male is black with white wing-patch and tail-coverts. The female is dark brown with faint streaking and a rusty rump. Young males are similar with a white wing-bar. The call is a harsh 'tsak tsak' and the song is a pretty little whistle, the male cocking its tail while singing.

WHITE'S THRUSH *Zoothera aurea* **28cm**

This large distinctive thrush breeds in north Asia and winters in the south. In the Philippines it is fairly common in Luzon, Palawan and Mindoro and reaches as far as Mindanao, occurring in forest from the lowlands to over 2,000m. In the early morning it feeds along forest trails and, when flushed, flies down the trail where it can be followed up. Otherwise, it is generally secretive and silent. It is olive-brown above and white below, with a buff band on the breast, all feathers having black and white crescent-shaped tips, with white tips to the outer tail and broad black and white bands on the underwing.

▲ *Male.*

BLUE ROCK-THRUSH *Monticola solitarius* 20cm

Breeding from Europe to Japan, this is a winter migrant to the tropics and is then fairly common and widespread in the Philippines, with a small population in the Batanes that breeds on cliffs near the shore. It prefers rocky slopes, buildings, quarries and road-cuts in forest, often perching prominently. The song is a delightful series of reedy whistles. The upperparts, throat and upper breast are greyish-blue, the wings and tail brownish-black. The lower breast and belly, typical of the race from north China and Japan, are chestnut. The female is greyish above mottled black and white, buffy below mottled black. Bill and feet are black.

▼ *Female.*

EYEBROWED THRUSH *Turdus obscurus* **22cm**

This medium-sized thrush is a winter visitor from its breeding grounds in northern Asia. It is not uncommon in the mountain forests, occurring in large flocks particularly when going to roost. Groups are often seen perched on dead snags at the forest edge or in fruiting trees where they feed on ripe berries. It calls in flight, a very high-pitched thin 'zeee'. The male is olive-brown with a rusty lower breast and flanks, and white belly. The chin, eyebrow and spot below the eye are white, and the head, throat and upper breast are grey. The female is paler and lacks the grey head.

ARCTIC WARBLER *Phylloscopus borealis* 12.5–13cm

A common migrant that breeds in Eurasia and Alaska and winters throughout southern Asia, this warbler is found in primary and secondary forest and wherever there are clumps of trees. It is usually solitary; it can be seen in the canopy looking for insects. The call is a sharp 'tzick'. It has a narrow, rather weak wing-bar (sometimes two), a long yellowish-white supercilium, uniform dark olive upperparts, brownish-olive flanks and dull creamy-white middle underparts and undertail-coverts. The lower mandible is yellowish with a dark tip.

LEMON-THROATED LEAF-WARBLER *Phylloscopus cebuensis* 12cm

This is one of three resident Philippine leaf-warblers, and is endemic to Luzon, Cebu and Negros where it occurs in lowland forests, singly or in mixed flocks. Its habits are similar to its relatives; the call is a distinct 'chip-chi-u'. The upperparts are dark olive-green, more greenish-yellow in tail and flight feathers, the underparts white, faintly streaked yellow. It possesses no wing-bar, a narrow yellow supercilium and pale yellow throat and undertail-coverts, although on Luzon the throat often appears white. The birds on Cebu and Negros have much brighter yellows.

PHILIPPINE LEAF-WARBLER *Phylloscopus olivaceus* 12cm

Resembles the Lemon-throated Leaf-warbler and is likewise endemic to the Philippines, but restricted to the southern islands, overlapping with Lemon-throated only on Negros, where it occurs at lower elevations. The upperparts are olive-green with an indistinct yellow wing-bar, while the underparts are white streaked with light yellow and with a pale yellow vent. The crown is grey with an olive wash, dark grey on the lores and behind the eye, and a yellowish eyebrow and eyering. The call is a high-pitched 'swee-iit chuua' on Mindanao, 'squiddly-chip' in the Sulus.

STREAKED REED-WARBLER *Acrocephalus sorghophilus* 13cm

This extremely rare warbler breeds in a restricted area in northern China (the millet fields of Liaoning province) and winters in the Philippines, seemingly in the wet marshes of Luzon where it prefers reedbeds or long grass near water. Formerly regular at Candaba Marshes and the Dalton Pass, this species is now rarely seen. A typical reed-warbler, it is golden-brown above streaked brown on the crown and back. It has a buffy supercilium bordered above by a black stripe on the side of the crown. The tail and wings are brown with buffy tips and edges. The throat and belly are white, the breast and flanks buffy.

TAWNY GRASSBIRD *Megalurus timoriensis* 21–24cm

This very large warbler is common throughout Mindanao, Luzon and the Visayas in open country with brush and thicket, plantations, and especially scrub-covered hillsides. Smaller and less obtrusive than the Striated Grassbird, it sits in or atop clumps of grass. The flight is weak and low over the ground. The song is a collection of ringing notes, the call a distinct 'tchak'. It has a rufous crown, brown upperparts with dark brown streaks, rufous-brown tail and unstreaked creamy-white underparts.

PHILIPPINE TAILORBIRD *Orthotomus castaneiceps* 13cm

Tailorbirds are bold, active little birds, found in bushes and thickets. Their name stems from the way they sew two sides of a leaf together to enclose their nests. Found in northern and central Luzon and on Panay, Negros, Cebu and related islands. The song is a variety of distinct loud liquid phrases. Found in forest, forest edge and overgrown thickets, it tends to feed near the ground but also climbs trees, particularly when agitated. In Luzon the crown and tail are chestnut, the back and rump yellowish-green. The throat and breast are grey streaked white, the belly white washed grey.

GREY-BACKED TAILORBIRD *Orthotomus derbianus* 12cm

This tailorbird is very similar and closely related to the Philippine Tailorbird, although it is a little smaller; the song and calls are indistinguishable, both species having a variety of loud, bubbly, liquid notes. It is endemic to central Luzon and found in lowland forest, forest edge, overgrown plantations and more cultivated country with thickets and bamboo groves. The main difference between the two species is that the Grey-backed has, indeed, a grey back and rump, greyish undertail-coverts and greyer underparts. Where it overlaps with the Philippine Tailorbird it has been found to have a smaller area of grey on the back.

WHITE-EARED TAILORBIRD *Orthotomus cinereiceps* 13cm

This little-known tailorbird is endemic to Basilan and western Mindanao where it is found in damp lowland forests up to 1,000m. It feeds on or near the ground in thick understorey and is detected by its loud song, which starts with a descending series that levels off to a continuous chatter of disyllabic 'dee-up' notes lasting a minute or two. Like all tailorbirds, the long bill and erect tail are diagnostic. The upperparts, flanks and undertail-coverts are olive-green, the top of the head is grey, the ear-coverts white, throat black, breast and belly grey.

YELLOW-BREASTED TAILORBIRD *Orthotomus samarensis* 12cm

Keeping close to the ground in the dense undergrowth of lowland forest, especially in vegetated gullies, this warbler is restricted to Bohol, Samar and Leyte but is closely related to the Black-headed Tailorbird *Orthotomus nigriceps* of Mindanao, their songs being very similar. The song begins as an explosive, rapid stuttering, given rapidly but descending and becoming longer and slower, and continuing as a monotonous series of single drawn-out notes. Males are olive-green above, yellow below, with a black head and throat with a small white patch on the chin. Females are similar but white on the throat and breast.

BRIGHT-HEADED CISTICOLA *Cisticola exilis* 9–11cm

This small warbler of dry grassy country, fields and gardens is not uncommon throughout the Philippines, although absent from Palawan. The call is a distinctive nasal buzzing 'speeee' often followed by a bell-like note, given from an exposed perch. When flushed it flies a short distance with a weak jerky flight before dropping into the grass. The nest is a ball of soft grass woven to a stem. The crown, lower back and rump are golden-rufous, the wings and rufous-tipped tail brown, the underparts cinnamon with a white belly. Non-breeding males and females have streaked crowns.

RUFOUS-TAILED JUNGLE-FLYCATCHER *Rhinomyias ruficauda* 15cm

Rhinomyias flycatchers are stolid, medium-sized, thick-set and heavy-billed, living in the substage of the forest. This bird is found in the southern Philippines from Samar and Bohol to Tawi-Tawi, where it is restricted to lowland primary and second-growth forests below 1,000m, but it also occurs in montane Borneo. Even when following a mixed feeding flock it tends to sit quietly 2–10m above ground for extended periods. The call is a quiet, high-pitched but rapid 1–3-note churr followed by a trill or whistle. Crown and back are brown, rump and tail rufous, wings dark brown, and underparts off-white with a faint brown breast-band.

WHITE-THROATED JUNGLE-FLYCATCHER *Rhinomyias albigularis* 16cm

This jungle flycatcher is very similar in appearance and habits to the Rufous-tailed, but is severely threatened with extinction, restricted as it is to the lowland forests of Negros and Panay. Recent fieldwork in the latter has shown that it is not as rare as once thought, but on Negros there is almost no habitat left for it. The song is an almost inaudible series of notes. The upperparts are brown to rufous-brown with a faint buff eyestripe; the tail is dark chestnut. The underparts are white with a brown breast-band.

GREY-STREAKED FLYCATCHER *Muscicapa griseisticta* 15cm

This medium-sized flycatcher, one of several congeners wintering in the Philippines, is by far the commonest, present from October to early May. Found along the forest edge or in the canopy, it perches on exposed branches from which it sallies out to catch insects. The upperparts are grey-brown, paler on the rump; the flight feathers are grey-brown and the underparts white, heavily streaked grey-brown. It has a white eye-ring and lores. The only bird it might be mistaken for is the Striped Flowerpecker *Dicaeum aeruginosum*, which is smaller and with different habits.

SNOWY-BROWED FLYCATCHER *Ficedula hyperythra* 11cm

Widespread from the Himalayas to the Moluccas, the Snowy-browed has numerous races including eight in the Philippines. Found in primary forest on most of the larger islands with mountains above 1,000m, it is common but unobtrusive, sitting low on a log or branch or hopping on the ground like a robin. The song is a very high-pitched phrase of 3–5 'zee' notes, one being lower. In general, birds have greyish-blue upperparts, an orange-rufous breast (variable between races), whitish belly and a small concealed white eyebrow. The tail varies from dark-brown to grey to chestnut. Females have olive-brown upperparts.

LITTLE SLATY FLYCATCHER *Ficedula basilanica* 12.5cm

This little flycatcher is endemic to the Phillipines and found on Mindanao, Basilan, Dinagat, Leyte and Samar. it inhabits the understorey of primary forest, in dark, shadowy places, often in the vicinity of streams. Inconspicuous, it is best located around dawn when the males give a simple whistled song, 'tsee-tsee-tsu' (recalling 'three-blind-mice'). The male is dark bluish-slate above with a narrow off-white stripe behind the eye. The underparts are off-white with a blue-grey wash on the breast and flanks. Females are rufous-brown above with brighter rufous lores and eyering and off-white underparts.

MOUNTAIN VERDITER-FLYCATCHER *Eumyias panayensis* 14cm

This medium-sized flycatcher inhabits the Moluccas, Sulawesi and, in the Philippines, Luzon, Mindoro, Negros, Panay and Mindanao. It occurs singly, in pairs and mixed flocks in montane forest and forest edge, and sings a catchy little song composed of a short series of 'swee' notes of various lengths and pitches. There are three races in the Philippines: in Luzon it is bright greenish-blue with greyish-blue wings, white belly and distinct black face and chin, while in Mindanao it is darker and duller, with a buffy belly and less distinct black on face.

PALAWAN FLYCATCHER *Ficedula platenae* 11cm

Secretive and uncommon, this small flycatcher is restricted to the lowland forests of Palawan. It lives singly or in pairs fairly close to the ground in the dense understorey, particularly among tangled rattan and old spreading vines. It is very site-faithful: the dense vegetation along the Balsahan River at Iwahig is a good place to see it. The call is a high-pitched but harsh 'tseee-tchaww'. Its other diagnostic feature is its bright orange-rufous tail. The upperparts are brown with a rufous tinge, darker on the wings; the cheeks, throat and breast are orange-buff, the belly white.

CRYPTIC FLYCATCHER *Ficedula crypta* 11cm

This small secretive flycatcher is endemic to Mindanao where it appears to be restricted to mid-montane forests between 700 and 1,200m, occurring between the Little Slaty Flycatcher *Ficedula basilanica* below it and the Snowy-browed above 1,000m, although some overlap occurs at the altitudinal boundaries. Cryptic is very similar to female Little Slaty, although the latter is not so rufous and has a larger bill. The upperparts are an olive-brown, becoming rustier towards the rump, the tail is rufous, and the underparts are pale, white on the belly with an indistinct olive-brown breast-band.

PALAWAN BLUE-FLYCATCHER *Cyornis lemprieri* 16cm

This medium-sized flycatcher is quite common but unobtrusive in the lower storeys of closed-canopy lowland forests in Palawan. The song is a soft whistle of 4–6 notes, initially descending then ascending. The male has dark blue upperparts including edges to wing and tail, with a lighter blue forehead and eye-stripe, and black lores and face; the breast is orange-rufous, paler on the throat, while the belly and undertail-coverts are white. The female has olive-brown upperparts and a chestnut tail. It also has a white broken eye-ring and malar stripe and black lores.

CITRINE CANARY-FLYCATCHER *Culicicapa helianthea* 11cm

This small perky flycatcher is found in Sulawesi and the Philippines where it is not uncommon in the mountain forests above 1,200m, although on some islands such as on Palawan it extends down to sea level. Mixed flocks often have a pair of these restless flycatchers which sit upright on the branch and fly out to catch insects before continuing with the flock. The call is a distinct, quickly given 'sweet-su sweet'. The upperparts are yellow-olive, the rump yellow and the underparts yellow with an olive wash on the breast. The wings and tail are brownish with yellow edges.

BLUE FANTAIL *Rhipidura superciliaris* 16cm

This bird is endemic to the lowland forests of Bohol, Samar, Leyte, Basilan and Mindanao up to 1,000m. Fairly secretive, it is found in the understorey of the forest either singly or in mixed flocks. The nest is built close to the ground concealed in a small shrub. The most obvious feature is its distinct call which is a rapid series of ascending clear whistles 'woo-woo-woo-...'. It has a long, often fanned-out tail, dark blue upperparts, paler blue supercilium, blue-edged black wings and tail, and greyish-blue underparts.

PIED FANTAIL *Rhipidura javanica* 19cm

Found also in the Greater Sundas and Malaysia, this typical fantail is very common throughout the Philippines, occurring singly, in pairs or small groups in mangrove and nipa swamps, plantations, bamboo groves, thickets and gardens. It hides in thick vegetation, actively hopping from branch to ground and back, constantly fanning its tail. The song and call are a series of raspy metallic notes. The upperparts are dark brown including the tail, which has broad white tips. The forehead is black, the eyebrow white. The underparts are white with a dark band across the upper breast.

BLACK-AND-CINNAMON FANTAIL *Rhipidura nigrocinnamomea* 16cm

This smart-looking fantail is endemic to the mountains of Mindanao where it occurs commonly in old forests over 1,500m. It is a prominent, active member of mixed flocks which move rapidly through the forest. The call is a single 'squeek', the song a distinct three-note whistle. The head is black with a white eyebrow meeting over the forehead, the back, rump, wing-coverts and tertials are cinnamon, the tail is rufous and the wings black edged cinnamon. In the northern mountains, such as in the Kitanglad range, birds are entirely cinnamon below, whilst in the south, such as on Mount Apo, the upper breast is white.

BLUE-HEADED FANTAIL *Rhipidura cyaniceps* M 18cm, F 15cm

Endemic to the Philippines, this is the common fantail of Luzon; another quite distinct (white-bellied) race occurs in the Western Visayas. It inhabits the understorey of forest from the lowlands to 2,000m, often as a prominent member and leader of mixed feeding parties, flying restlessly from branch to branch, flashing its tail. The head, upper back, throat and breast are greyish-blue, the lores and ear-coverts black, the lower back, rump, belly and outer tail rufous, the centre of tail and wings black. In Luzon the call is a metallic 'chik chik' or staccato 'chip-chip-chip'; in the Visayas it is quite different.

BLACK-NAPED MONARCH *Hypothymis azurea* 15cm

Widespread from India to the Moluccas, this flycatcher is common throughout the Philippine lowlands in forest, dense thickets and wooded valleys. Lively and inquisitive, it is found alone, in pairs and as a member of mixed feeding flocks. The call is a scratchy 'chi-chweet' and the song a series of clear whistles, 'whit-whit whit', the notes not as long or forced as the Rufous Paradise-flycatcher. The male is bright azure, darker on the back, with black wings and tail, grey-white belly and white vent; forehead, nape and throat-crescent are black. Females lack the black and have brown backs.

RUFOUS PARADISE-FLYCATCHER *Terpsiphone cinnamomea* 21cm

One of the most beautiful birds in the Philippines, found on most islands but not Palawan, and not yet Bohol or Leyte, this is an uncommon resident in primary and secondary lowland forest up to 1,000m, appearing to be one of the leaders of mixed feeding flocks. The male is dark orange-red with an exceptionally long tail and large sky-blue eye-wattles, blue-black bill and bluish feet. The female is duller with a shorter tail. Birds in the south also have shorter tails and are less orange. The call is a strident, forceful, repeated 'swheek-swheek-swheek-swheek'.

BLUE PARADISE-FLYCATCHER
Terpsiphone cyanescens **21cm**

This large flycatcher is endemic to Palawan and its satellites, and is found in the thickets and understorey of lowland primary and secondary forest. Its presence is indicated by its high-pitched staccato trill, rising in intensity and lasting about four seconds. It is short-crested and has an obvious large head. The male is a powdery grey-blue with black flight feathers, lores, chin and narrow black throat-crescent. The female has a blue head and throat, becoming grey-blue on breast and white on belly. The back, wings and tail are brown.

YELLOW-BELLIED WHISTLER *Pachycephala philippinensis* **15cm**

Whistlers are robust, forest-dwelling, chiefly Australasian relatives of the flycatchers with big heads and shrike-like bills. Of the four species in the Philippines, the Yellow-bellied is the most widespread although not found in Palawan, Mindoro or the Western Visayas. It inhabits the canopy of old forests in the lowlands, although in the south it reaches at least 1,800m. A secretive bird, it is heard more often than seen but joins mixed flocks. The call is a plaintive 1–2-note whistle. The upperparts are yellowish-olive, top of head olive-brown, throat white and breast and belly yellow.

WHITE-VENTED WHISTLER *Pachycephala homeyeri* 16cm

This whistler is found on the Sulu islands, Mindanao and the Western Visayas, but is not quite endemic to the Philippines owing to a population on the islands of Sipadan and Siamil off Sabah. It lives alone or in pairs in the understorey of primary and secondary forest from the lowlands to over 2,000m, where it is usually detected by its call, a short series of 1–9 whistles with stress on one note, usually the penultimate. It is fairly drab, being reddish-brown above and whitish below with a grey wash on the breast. It could be confused with White-throated Jungle-Flycatcher.

GREY WAGTAIL *Motacilla cinerea* 18cm

This widespread Old World species is a common winter visitor to the Philippines from September to May. It is found in open country along streams and rivers, also creeks in forest and built-up areas. It feeds on the ground, flying low up the stream when flushed, and vigorously wagging its tail when landing. The call, given in flight, is a sharp high-pitched 'stit-it'. It is distinguished from the Eastern Yellow Wagtail *Motacilla tschutschensis* by its grey mantle, white wing-bar, yellow-green rump and longer tail. The underparts are yellow; the throat is white in winter, black when breeding.

WHITE WAGTAIL *Motacilla alba* 19cm

The White Wagtail is an uncommon to rare winter visitor to the Philippines, mostly to the Batanes and northern Luzon. All records are of the race ocularis, although lugens has been claimed. It is found in open, generally wet areas; the stony banks of larger rivers are favoured sites. Forecrown, face and underparts are white except for a black breast-band and a black eye-line. It has a black hindcrown, grey upperparts, white wing-coverts, dark grey wings with white edges and long black tail with white outer tail feathers. The call is a hard 'chissik' given in flight.

PADDYFIELD PIPIT *Anthus rufulus* 16cm

Pipits and wagtails are small terrestrial birds that stand upright and walk very actively. Paddyfield Pipit is medium-sized with long strong legs, found in open grassland, meadows, dried paddies and grassy hillsides. It likes to stand at a prominent site on the ground looking for prey. If flushed, it runs fast with head up, or flies a short distance then runs into some furrow. The flight is undulating and the call, given regularly in flight, is a 'chup'. The species has a pale supercilium and white outer tail feathers, and is streaked dark brown and sandy-buff above and light buff below with faint streaking on the breast.

WHITE-BREASTED WOODSWALLOW *Artamus leucorynchus* 19cm

This woodswallow, found from Malaysia to Australia, is common throughout the islands in open country with scattered trees, coconut groves, clearings and coastal areas up to 1,500m or more. It is a graceful flier, often sailing motionless in the air for periods of time. It perches motionless in small close groups on telegraph wires and dead branches, sallying out either to catch insects or to mob some large bird passing by. It builds cup-shaped nests in broken branches. The upperparts, throat, tail and wings are slate-grey and the belly and rump are white. The call is a shrill 'ik-kik-kik-kik' given in flight.

BROWN SHRIKE *Lanius cristatus* 19cm

This abundant winter visitor occurs throughout the country from sea level to the mountains, and from the forest edge to the centre of Manila, in all types of habitat except forest proper. It is present from mid-September to May and easily noticed by its harsh, chattering 'che-che-che'. It sits on fence posts, telegraph wires and bushes and preys aggressively on large insects and small birds. The upperparts are brown with some rufous on the rump and tail-coverts. The underparts are white washed with rufous-buff on belly and flanks. There is a black band through the eye and an ill-defined white eyebrow. The crown shows a varying degree of grey.

LONG-TAILED SHRIKE *Lanius schach* 24cm

Noisy, aggressive and territorial, pairs of this very smart bird are common throughout the Philippines except Palawan, in dry open grassy country with scattered woody vegetation at least to 1,800m. It perches atop trees and bushes, swooping on small birds and lizards. The calls are scolding and often mimic other birds. It has a powerful black hooked bill, black head and hindneck, light grey mantle and upper back, light rufous rump, flanks and belly, white throat and breast, and long black tail and short black wings, both with white markings. Immatures are duller with barred flanks and back and greyer head.

MOUNTAIN SHRIKE *Lanius validirostris* 20cm

This uncommon shrike is endemic to the mountains of the Philippines and divides into three races found, respectively, in Luzon, Mindoro and Mindanao. It mainly occurs above 1,250m, becoming commoner in the high-altitude oak and pine forests where it is seen at the forest edge either singly or in pairs. The call is a loud harsh thrush-like whistle. It is a uniform slaty-grey from crown to rump, with blackish-brown wings and tail, white underparts with rufous on the flanks, and a black mask through the eye. It lacks the white eyebrow and forecrown of the Brown Shrike.

ASIAN GLOSSY STARLING *Aplonis panayensis* 20cm

Common throughout the Philippines in towns, villages, cultivated areas and forest edge, ranging from sea level to 1,000m, this starling is often seen flying in flocks around cities, especially in the south, looking for fruiting trees. It perches in groups at the tops of trees, and nests in colonies in holes or forks of branches in trees, bamboo stands and even walls. The adult is entirely black with a strong green gloss on the back. The bill and feet are black. Young birds are blackish-brown above and white below, heavily streaked black. The call is a loud 'seeup' with a variety of metallic clicking calls.

CRESTED MYNA *Acridotheres cristatellus* 25cm

Originally introduced from China last century, this large gregarious starling is patchily distributed in the lowlands around Luzon and has been seen on Negros. It inhabits open country, cultivated land, plantations and drained fishponds, commonly being seen beside main roads perched on telephone wires. The nest, made of grass, is placed in the crevices of trees and buildings. At rest, the bird appears black with an obvious crest on the forecrown and yellow bill and legs. There is a white tip to the tail, the undertail-coverts show white scaling and, in flight, distinctive white wing-patches are seen.

COLETO *Sarcops calvus* 29cm

A variety of clicks, tinkles and whistles, like an orchestra warming up, announces the presence of the noisy Coleto, a common long-tailed myna, endemic to the Philippines (but not on Palawan). It has a large bare pinkish wattle around each eye that covers most of the head, but the plumage is mostly black except the mantle, back, rump, undertail-coverts and flanks, which are silvery grey. In the southern islands the upper back is black. Usually in small groups of two to five birds, often perched on bare branches at the forest edge, in secondary growth and cultivations, it typically eats fruit and nests in tree holes or atop coconut stumps.

HILL MYNA *Gracula religiosa* 27cm

This is the well-known 'talking' Asian myna, a popular cagebird, much persecuted in the wild. In the Philippines it is confined to the Palawan group where it is now very uncommon. It is bulky, noisy and obvious, sitting atop a tall tree, usually near forest, often imitating other birds such as Koels or Palawan Peacock-pheasants. It is all black with a greenish or blue wash and a white bar on the wing, a bare yellow patch below the eye and a yellow skinfold from the eye to the nape. The bill is orange-red with a yellow tip and the legs are yellow.

▲ *Male.*   ▼ *Female.*

OLIVE-BACKED SUNBIRD *Cinnyris jugularis* 11cm

The commonest sunbird, widespread from South-East Asia to the South Pacific, present in the Philippines in more open country with scattered bushes, plantations, mangroves and gardens. It is usually seen in pairs feeding in flowering trees and bushes; very active, it also sits atop bushes calling loudly 'che-wheeet'. The male is olive-green above and bright yellow below with a black throat that has a metallic blue iridescence when breeding. The tail is black with white spots on the outer feathers. The female is similar but the entire underparts are dull yellow. Some races are bright orange on the upper breast.

▲ Male.   ▲ Female.

PURPLE-THROATED SUNBIRD *Leptocoma sperata* 10cm

Fairly common in five races throughout the Philippines, this small Asian nectar-eater is found, usually in pairs, in the lowlands up to 1,500m in secondary forest, plantations, open country, mangroves and gardens, and is often to be seen at flowering coconut trees. It weaves a small pendant nest from a branch on a bush. It has a metallic green crown, shoulders and lower back, velvet black face, metallic purple throat, black tail and (red-edged) wings, scarlet breast and olive-yellow belly to undertail. In western Mindanao and the Sulus the breast is bright yellow; in northern Luzon the mantle is black.

GREY-HOODED SUNBIRD *Aethopyga primigenia* 11cm

This sunbird is endemic to Mindanao and is restricted to mountainous regions between 1,000 and 1,500m where, however, it is common. It inhabits primary forest and forest edge, occurring singly, in pairs or mixed flocks. It feeds on flowering trees and shrubs and is particularly fond of the flowering banana. The head and upper breast are grey, forming a hood. The rest of the upperparts including the wings and tail are olive-green, the rump yellow. The breast and belly are white with bright yellow flanks and undertail-coverts. The call includes metallic 'pink pink' notes in a level or ascending series.

METALLIC-WINGED SUNBIRD *Aethopyga pulcherrima* 10cm

This was once known as the Mountain Sunbird because on Luzon it occurs in mountains and foothills, but on Mindanao it inhabits lowlands and mid-levels. It is a bird of forest and edges and, in the south, of plantations and coconut groves. The long curved bill and very short bluish-green tail give it a distinctive shape. The male is olive-green above with a metallic blue-green forecrown and wing-coverts; throat and breast are bright yellow. Females are greyish-yellow. Both sexes have yellow rumps. The call is a high-pitched metallic 'zip zip', often forming a trill.

CRIMSON SUNBIRD *Aethopyga siparaja* 13cm

This widespread Asian beauty is restricted in the Philippines to Cebu, Negros and Panay and some smaller islands. It is found in forest edges, cultivations, gardens and open country with scattered trees in the lowlands. The male is crimson on the crown, face, neck and upper back and a lighter crimson from chin to breast. The belly and wings are black, the rump yellow. The forecrown and malar stripe are iridescent purple, as are the edges to the black tail. The female has a dark olive-green head, back and rump, greyer green underparts and red-washed dark brown wings and blackish tail.

PALAWAN FLOWERPECKER *Prionochilus plateni* 9cm

This small flowerpecker is common and endemic to Palawan where it lives in a variety of habitats: forest, forest edge, plantations, cultivations and gardens. The call, a high-pitched 'seeep seeep', is typical of its family. The male is slaty-blue above including the sides of the head, has a yellow rump and an orange patch on the crown. The distinct white moustachial stripe is diagnostic. The underparts are yellow with a red spot in mid-breast. The female is olive-green above with a yellow rump and less distinct moustachial stripe, and yellow below with a white throat.

OLIVE-BACKED FLOWERPECKER *Prionochilus olivaceus* 9cm

More associated with primary forest than most flowerpeckers, this bird is found deep in the understorey, often close to the ground, and is also seen travelling with mixed feeding flocks. Found on Mindanao, Basilan, Luzon and the Eastern Visayas, in the lowlands below 1,000m, it has a distinct chunky bill which is diagnostic even in silhouette. The upperparts are bright olive-green, the wings and tail being browner. The underparts are white in males, greyer in females; Luzon males have nearly black lores, sides to throat and breast; Mindanao males are pale grey instead of black. The call is a loud single 'peeit'.

OLIVE-CAPPED FLOWERPECKER *Dicaeum nigrilore* 10cm

This common but little-known flowerpecker is restricted to the mid-mountain and mossy forests of Mindanao above 1,000m where it occurs in monospecific or mixed flocks when feeding in flowering trees. The call is a high-pitched scratchy 'ziti-ziti-' repeated 10–15 times. It has a very long curved bill for a flowerpecker, but is not unlike a miniature version of the Black-masked White-eye *Lophozosterops goodfellowi*, with a bright olive-green top of head, rump and uppertail-coverts, black lores and line below eye, brown mantle, yellow flanks and vent, white breast and abdomen, and ash-grey throat.

BUZZING FLOWERPECKER *Dicaeum hypoleucum* 9cm

The Buzzing Flowerpecker is so named for its distinctive 'bzzzeeep bzzzeeep' call, often given in a series. Very plain save the long thin decurved bill. The upperparts are olive; the underparts on Luzon birds have a greyish wash with a little yellow from the mid-breast to undertail-coverts. In the south, the underparts are more distinctly white. It is endemic to the Philippines, found throughout except Palawan, in deep forest and forest edge in the lowlands up to around 1,500m. Often feeds on fig fruits growing near the base of the trees.

▲ *Southern race.*
▼ *Luzon race.*

SCARLET-COLLARED FLOWERPECKER *Dicaeum retrocinctum* 10cm

This bird is endemic to Mindoro, usually occurring below 1,000m in the canopy and edge of forest and in open country with scattered trees. The call is a series of notes similar to striking two stones together plus a high-pitched 'zeet zeet zeet'. It is closely related to the Red-keeled Flowerpecker *Dicaeum australe* but has a longer, more slender curved bill. The upperparts are black with a slate-blue gloss and scarlet collar across the upper back. The throat and upper breast are black with a small red patch on the throat, the belly is white with a scarlet ventral stripe bordered with black.

ORANGE-BELLIED FLOWERPECKER *Dicaeum trigonostigma* 9cm

This widespread Asian flowerpecker has no fewer than eleven races in the Philippines, many quite distinct from mainland birds and none having the latters' orange rump (only in Luzon is the rump yellow). Males, generally, are yellow below, sometimes with an orange patch, and blue above with an orange patch on the upper back. Some races have a grey throat and olive-green rump. Females, which are olive-green below, are yellow from the mid-breast to undertail-coverts. The species inhabits forest edge, secondary scrub and plantations. Calls include a high-pitched 'zeeeep zeeeep' and metallic clicks.

LOWLAND WHITE-EYE *Zosterops meyeni* 10cm

White-eyes get their name from the white feathers around the eye. They are small, active, noisy birds that form loose flocks feeding on insects and small fruits. The Lowland White-eye is endemic to Luzon and the northern islands; it is found in more open country with scattered trees, plantations, gardens and vegetation near marshland. It has black lores, a broad white eye-ring and light yellow forehead, throat and undertail-coverts. The upperparts are a yellowish olive-green and the breast and belly are white. The call is a series of high-pitched twittering 'swit-zee' notes.

MOUNTAIN WHITE-EYE *Zosterops montanus* 11cm

Found throughout Indonesia and the Philippines in many different races, this species is found in the mountains over 1,000m and replaces lowland forms. Its loose, often quite large flocks move actively through the forest and forest edge feeding on tiny fruits and insects; on Mindanao in particular it is abundant (the commonest bird) above 1,500m. The upperparts are olive-green, yellower on the forehead, yellow on throat and undertail-coverts and white on belly and breast. It differs from the Lowland White-eye in being less yellowish above, with a darker, more extensive yellow on the throat and a black spot interrupting the front of the white eye-ring.

EURASIAN TREE SPARROW *Passer montanus* 13cm

Although probably originating from China, this small ubiquitous familiar sparrow now probably occurs on every inhabited island in the Philippines, having found its niche in the relatively recent habitats created by man. It appears to breed throughout the year, the nest being an untidy ball of grass placed in the eaves and crevices of buildings. It is also a pest, descending in thousands when the rice fields are being harvested. Its distinctive features are a chestnut cap, black chin and white cheeks with black spot in the centre. The upperparts are brown with blackish streaks, the underparts dirty pale buff.

RED-EARED PARROT-FINCH *Erythrura coloria* 10.5cm

This small, little-known finch with a wedge-shaped tail was first discovered in 1960; it is endemic to the mountains of Mindanao over 1,000m. It lives in forest and forest edge, keeping low in the dense understorey, particularly the blackened branches of old palms, quietly moving about in search of seeds and fruits. When flushed, it flies fast and direct like a green bullet, giving a short sharp 'prrrt'; it also has a sharp repeated 'tik'. The body is mostly dark green with red uppertail-coverts and central tail feathers; the outer feathers are reddish-brown. The forehead and face are blue, the ear-coverts and sides of face red.

SCALY-BREASTED MUNIA *Lonchura punctulata* 11cm

Found from India through to South China and Sulawesi, this estrildine finch is a fairly common resident on Luzon, Palawan, Mindanao and the Western Visayas. It lives in open country with scattered trees, marshlands, paddyfields, cultivations and gardens, travelling in compact flocks and often mixing with other munias. The nest is globular, made of grass. The upperparts are brown with buffy shaft-streaks; the underparts are brown and white with scale-like markings, and the tail is brown with yellowish edges. The diagnostic rich dark brown face and throat are obvious in the field. The call is a disyllabic 'ke-tee'.

CHESTNUT MUNIA *Lonchura atricapilla* 11cm

This munia is abundant from India to China and all of South-East Asia including the Philippines, where it was once the national bird, known as the 'Maya' or ricebird, living in open grasslands, rice paddies and marshland. Often forming large compact flocks, it can be a pest on rice crops. It breeds in colonies, attaching its globular grass-woven nest to tall grasses and shrubs. The back, wings, tail and sides of belly are chestnut whilst the head, neck, throat, upper breast, centre of belly and undertail-coverts are black. The young bird is reddish-brown above, buffy below, with no black head.

WHITE-BELLIED MUNIA *Lonchura leucogastra* 11cm

This familiar little finch occurs in Malaysia, Sumatra, Borneo and throughout the Philippines in a variety of habitats including forest understorey, forest edge, marshlands, open grassland and paddyfields, attaching its globe-shaped nest to a branch. It lives in flocks often with other munias, and in flight keeps together in tight units. It is brown above with white shaft-streaks on the neck, back and wing-coverts. The brown tail is bordered with yellow. The head, throat, chest and flanks are blackish-brown; the belly and mid-breast are white with a few black marks. The call is a variety of soft tinkling 'cheeps'.

WHITE-CHEEKED BULLFINCH *Pyrrhula leucogenis* 16cm

This medium-sized stout-billed finch is endemic to the Philippines where it is restricted to certain mountains on Mindanao, Panay and Luzon. Found singly or in small flocks, it is an uncommon bird of the forest canopy, edge and clearings with scattered trees, being commoner in the mossy oak forests of higher altitudes. The call is a clear but soft and melodious 'pee-yoo'. The face, crown, wings and tail are black with a purplish gloss. The mantle and underparts are brown except the mid-belly which is white and undertail-coverts which are buff. The white cheek-patch is diagnostic.

INDEX

Accipiter virgatus 66
Aceros leucocephalus 84
    waldeni 85
Acridotheres cristatellus 129
Acrocephalus sorghophilus 112
Actenoides hombroni 80
    lindsayi 80
Actitis hypoleucos 47
Aegithina tiphia 93
Aethopyga primigenia 132
    pulcherrima 133
    siparaja 133
Alcedo argentata 76
    atthis 75
    cyanopectus 75
Amaurornis olivacea 39
    phoenicurus 39
Anas clypeata 30
    luzonica 31
    querquedula 29
Anhinga melanogaster 21
Anous minutus 52
    stolidus 52
Anthus rufulus 126
Aplonis panayensis 129
Ardea alba 23
    purpurea 23
    sumatrana 22
Arenaria interpres 48
Artamus leucorynchus 127
Aythya fuligula 30

Babbler, Chestnut-faced 104
    Flame-templed 102
    Melodious 104
Balicassiao 97
Barbet, Coppersmith 87
Batrachostomus septimus 72
Bee-eater, Blue-tailed 82
    Blue-throated 81
Besra 66
Bittern, Cinnamon 28
    Schrenck's 28
    Yellow 27
Bleeding-heart, Luzon 62
    Mindanao 62
    Mindoro 63
Blue-flycatcher, Palawan 120
Bolbopsittacus lunulatus 66

Booby, Brown 20
Brachypteryx montana 105
Broadbill, Mindanao Wattled 89
    Visayan Wattled 89
Brown-dove, Amethyst 55
    White-eared 55
Bubo philippensis 70
Bubulcus coromandus 24
Buceros hydrocorax 86
Bulbul, Black-headed 94
    Grey-cheeked 95
    Philippine 95
    Sulphur-bellied 95
    Yellow-vented 94
    Yellowish 96
Bullfinch, White-cheeked 140
Bushchat, Pied 108
Bush-hen, Plain 39
Butorides striata 26
Buttonquail, Spotted 38

Cacatua haematuropygia 64
Cacomantis merulinus 67
    sepulcralis 67
    variolosus 67
Calidris acuminata 48
    subminuta 48
Caloenas nicobarica 63
Canary-flycatcher, Citrine 120
Caprimulgus affinis 73
Centropus bengalensis 68
    viridis 68
Ceyx lepidus 76
    melanurus 77
Chalcophaps indica 61
Charadrius alexandrinus 42
    dubius 42
    leschenaultii 43
    mongolus 43
    peronii 42
Chlidonias hybrida 53
    leucopterus 53
Ciconia episcopus 29
Cinnyris jugularis 131
Cisticola, Bright-headed 115
Cisticola exilis 115
Cockatoo, Philippine 64
Colasisi 65
Coleto 130

Collared-dove, Island 60
Collocalia esculenta 74
Columba vitiensis 58
Copsychus cebuensis 107
    luzoniensis 107
    saularis 106
Corvus macrorhynchos 100
Coturnix chinensis 36
Coucal, Lesser 68
    Philippine 68
Crake, Slaty-legged 38
    White-browed 39
Criniger bres 95
Crow, Large-billed 100
Cuckoo, Brush 67
    Plaintive 67
    Rusty-breasted 67
Cuckoo-dove, Philippine 59
Cuculus pectoralis 66
Culicicapa helianthea 120
Cyornis lemprieri 120

Darter, Oriental 21
Dendrocopus maculatus 88
Dendrocygna arcuata 31
Dicaeum aeruginosum 117
    australe 136
    hypoleucum 135
    nigrilore 135
    retrocinctum 136
    trigonostigma 136
Dicrurus balicassius 97
    hottentottus 96
Dollarbird 82
Dove, Emerald 61
    Spotted 60
    Zebra 60
Dowitcher, Asian 44
Drongo, Hair-crested 96
Dryocopus javensis 88
Duck, Philippine 31
    Tufted 30
Ducula aenea 58
    bicolor 59

Eagle, Philippine 34
    Rufous-bellied 34
Eagle-owl, Philippine 70
Egret, Chinese 26

Eastern Cattle 24
Great 23
Little 24
Egretta eulophotes 26
  garzetta 24
  sacra 25
Erythrura coloria 138
Esacus magnirostris 49
Eudynamys scolopacea 68
Eumyias panayensis 118
Eurylaimus samarensis 89
  steerii 89
Eurystomus orientalis 82

Fairy-bluebird, Asian 99
  Philippine 100
Falconet, Philippine 35
Fantail, Black-and-cinnamon 122
  Blue 121
  Blue-headed 122
  Pied 121
Ficedula basilanica 119
  crypta 119
  hyperythra 117
  platenae 119
Flowerpecker, Buzzing 135
  Olive-backed 134
  Olive-capped 135
  Orange-bellied 136
  Palawan 134
  Red-keeled 136
  Scarlet-collared 136
  Striped 117
Flycatcher, Cryptic 119
  Grey-streaked 117
  Little Slaty 118
  Palawan 119
  Snowy-browed 117
Fregata ariel 22
Frigatebird, Lesser 22
Frogmouth, Philippine 72
Fruit-dove, Black-chinned 57
  Flame-breasted 57
  Yellow-breasted 56

Gallicolumba crinigera 62
  luzonica 62
  platenae 63
Gallinula chloropus 41

Gallus gallus 37
Garganey 29
Gelochelidon nilotica 51
Geopelia striata 60
Glareola maldivarum 50
Godwit, Bar-tailed 44
  Black-tailed 44
Gracula religiosa 130
Grassbird, Tawny 113
Grass-owl, Eastern 71
Grebe, Little 20
Green-pigeon, Ashy-headed 54
  Pink-necked 54
  Thick-billed 54
Greenshank, Common 44
Ground-babbler, Streaked 103
Guaiabero 66
Gull, Black-headed 51

Halcyon coromanda 78
  smyrnensis 79
Haliaeetus leucogaster 33
Haliastur indus 32
Hawk-cuckoo, Philippine 66
Hawk-eagle, Philippine 35
Hemiprocne comata 74
Heron, Great-billed 22
  Little 26
  Purple 23
Hieraaetus kienerii 34
Himantopus himantopus 50
Hirundo rustica 91
  tahitica 92
Hornbill, Luzon 83
  Rufous 86
  Walden's 85
  Writhed 84
Hypothymis azurea 123

Imperial-pigeon, Green 58
  Pied 59
Iora, Common 93
Irena cyanogastra 100
Irena puella 99
Ixobrychus cinnamomeus 28
  eurythmus 28
  sinensis 27
Ixos everetti 96
  palawanensis 96
  philippinus 96

Jungle-flycatcher, Rufous-tailed 116
  White-throated 116, 125
Junglefowl, Red 37

Kingfisher, Blue-capped 80
  Collared 77
  Common 75
  Indigo-banded 75
  Philippine Dwarf 77
  Ruddy 78
  Rufous-lored 81
  Silvery 76
  Stork-billed 78
  Variable Dwarf 76
  White-throated 79
Kite, Brahminy 32
Koel, Asian 68

Lalage melanoleuca 93
  nigra 93
Lanius cristatus 127
  schach 128
  validirostris 128
Larus ridibundus 51
Leaf-warbler,
  Lemon-throated 111
  Philippine 112
Leptocoma sperata 132
Limnodromus semipalmatus 44
Limosa lapponica 44
  limosa 44
Lonchura atricapilla 139
  leucogastra 140
  punctulata 139
Lophozosterops goodfellowi 135
Loriculus philippensis 65
Luscinia calliope 106

Macronous striaticeps 105
Macropygia tenuirostris 59
Magpie-robin, Oriental 106
Malacopteron palawanense 104
Malkoha, Chestnut-breasted 67
Megalaima haemacephala 87
Megalurus timoriensis 113
Megapodius cumingii 36
Merops philippinus 82
  viridis 81
Microhierax erythrogenys 35

Mimizuku gurneyi 71
Monarch, Black-naped 123
Monticola solitarius 109
Moorhen, Common 41
Motacilla alba 126
   cinerea 125
   tschutschensis 125
Mulleripicus funebris 87
Munia, Chestnut 139
   Scaly-breasted 139
   White-bellied 140
Muscicapa griseisticta 117
Myna, Crested 129
   Hill 130

Night-heron, Black-crowned 27
   Rufous 27
Nightjar, Savanna 73
Noddy, Brown 52
Numenius phaeopus 43
Nuthatch, Sulphur-billed 101
   Velvet-fronted 101
Nycticorax caledonicus 27
   nycticorax 27

Oriole, Black-naped 98
Oriolus chinensis 98
Orthotomus castaneiceps 113
   cinereiceps 114
   derbianus 114
   nigriceps 115
   samarensis 115
Osprey 32
Otus megalotis 69

Pachycephala homeyeri 125
   philippinensis 124
Pandion haliaetus 32
Paradise-flycatcher, Blue 124
   Rufous 123
Pardaliparus elegans 101
Parrot, Blue-naped 65
Parrot-finch, Red-eared 138
Passer montanus 138
Peacock-pheasant, Palawan 37
Pelargopsis capensis 78
Penelopides manillae 83
Phapitreron amethystina 55
   leucotis 55
Phylloscopus borealis 111

   cebuensis 111
   olivaceus 112
Pigeon, Metallic 58
   Nicobar 63
Pipit, Paddyfield 126
Pithecophaga jefferyi 34
Pitta, Black-headed 90
   Green-breasted 90
   Hooded 90
   Steere's 91
   Whiskered 89
Pitta kochi 89
   sordida 90
   steerii 91
Plover, Kentish 42
   Little Ringed 42
   Malaysian 42
   Pacific Golden 41
Pluvialis fulva 41
Polyplectron napoleonis 37
Porphyrio pulverulentus 40
Porzana cinerea 39
Pratincole, Oriental 50
Prioniturus platenae 64
Prionochilus olivaceus 134
   plateni 134
Ptilinopus marchei 57
   occipitalis 56
Ptilocichla mindanensis 103
Ptilonopus leclancheri 57
Pycnonotus atriceps 94
   goiavier 94
Pyrrhula leucogenis 140

Quail, Blue-breasted 36

Racquet-tail, Blue-crowned 64
Rallina eurizonoides 38
Redshank, Common 45
   Spotted 45
Reed-warbler, Streaked 112
Reef-egret, Pacific 25
Rhabdornis,
   Stripe-headed 102
Rhabdornis inornatus 102
Rhinomyias albigularis 116
   ruficauda 116
Rhipidura cyaniceps 122
   javanica 121
   nigrocinnamomea 122

   superciliaris 121
Rock-thrush, Blue 109
Roller, Broad-billed 82
Rostratula benghalensis 49
Rubythroat, Siberian 106

Sandpiper, Common 47
   Green 46
   Marsh 44
   Sharp-tailed 48
   Wood 46
Sand-plover, Greater 43
   Lesser 43
Sarcops calvus 130
Saxicola caprata 108
Scolopax bukidnonensis 46
Scops-owl, Giant 71
   Philippine 69
Scrubfowl, Tabon 36
Sea-eagle, White-bellied 33
Serpent-eagle, Crested 33
   Philippine 33
Shama, Cebu Black 107
   White-browed 107
Shortwing, White-browed 105
Shoveler, Northern 30
Shrike, Brown 127
   Long-tailed 128
   Mountain 128
Sitta frontalis 101
   oenochlamys 101
Snipe, Greater Painted 49
Sparrow, Eurasian Tree 138
Spilornis cheela 33
   holospilus 33
Spizaetus philippensis 35
Stachyris speciosa 102
   whiteheadi 104
Starling, Asian Glossy 129
Sterna sumatrana 52
Stilt, Black-winged 50
Stint, Long-toed 48
Stork, Woolly-necked 29
Streptopelia bitorquata 60
   chinensis 60
Sula leucogaster 20
Sunbird, Crimson 133
   Grey-hooded 132
   Metallic-winged 133
   Olive-backed 131

Purple-throated 132
Swallow, Barn 91
    House 92
Swamp-hen, Philippine 40
Swiftlet, Glossy 74

Tachybaptus ruficollis 20
Tailorbird, Black-headed 115
    Grey-backed 114
    Philippine 113
    White-eared 114
    Yellow-breasted 115
Tanygnathus lucionensis 65
Tattler, Grey-tailed 47
Tern, Black-naped 52
    Gull-billed 51
    Whiskered 53
    White-winged Black 53
Terpsiphone cinnamomea 123
    cyanescens 124
Thick-knee, Beach 49
Thrush, Eyebrowed 110
    White's 108
Tit, Elegant 101
Tit-babbler, Brown 105

Todiramphus chloris 77
    winchelli 81
Treeswift, Whiskered 74
Treron curvirostra 54
    phayrei 54
    vernans 54
Triller, Black-and-white 93
    Pied 93
Tringa brevipes 47
    erythropus 45
    glareola 46
    nebularia 44
    ochropus 46
    stagnatilis 44
    totanus 45
Turdus obscurus 110
Turnix ocellatus 38
Turnstone, Ruddy 48
Tyto longimembris 71

Verditer-flycatcher,
    Mountain 118

Wagtail, Eastern Yellow 125
    Grey 125

    White 126
Warbler, Arctic 111
Waterhen, White-breasted 39
Whimbrel 43
Whistler, White-vented 125
    Yellow-bellied 124
Whistling-duck, Wandering 31
White-eye, Black-masked 135
    Lowland 137
    Mountain 137
Woodcock, Bukidnon 46
Wood-kingfisher, Spotted 80
Woodpecker, Philippine
    Pygmy 88
    Sooty 87
    White-bellied 88
Woodswallow, White-
    breasted 127

Zanclostomus curvirostris 67
Zoothera aurea 108
Zosterops meyeni 137
    montanus 137

PHOTO CREDITS

D. Allen: 42T, 59B, 67T, 68T, 76T, 78T, 81T, 93T, 96T, 125T, 132TL, 132TR, 135T, 140T; Alain Compost: 29T, 128T; David M. Cottridge: 29B, 53B; Chew Yen Fook: 27B, 39T, 39B, 68B; J. C. Gonzalez: 38T, 57T, 69B, 77T, 88B, 97B, 101T, 102B, 104T, 112B, 114T, 122B, 123T, 123B, 124B, 133B, 134B, 135B, 136T; A. Greensmith: 35T, 57B, 74B, 107B, 118B, 134T; Simon Harrap: 22B, 25, 46B, 49B, 52T, 53T, 55B, 66B, 71B, 83, 87B, 102T, 103, 112T, 117B, 118T, 128B; Jon Hornbuckle: 49T, 64T, 77B, 80T, 85, 87T, 88T, 89T, 91T, 95T, 95B, 97T, 101B, 104B, 105T, 105B, 106T, 107T, 111T, 111B, 113T, 113B, 114B, 115T, 116T, 116B, 119T, 119B, 120T, 121T, 121B, 122T, 124T, 125B, 132B, 133T, 135C, 136B, 138B; Yasuyuki Makino: 109T, 109B; Peter Morris: 20T, 23B, 26T, 33BL, 41B, 43B, 44T, 46T, 47T, 51B, 63T, 69T, 73, 80B, 100T, 140B; T C Nature: 36T, 139T; Bill Simpson: 26B, 28B, 34B, 47B, 72, 89B, 120T; Sam Stier: 66T; Morten Strange: 42B, 67B, 115B; Tony Tilford: 93B; David Tipling: Windrush Photos: 32T, 108B; Ray Tipper: 48B